Books on scientific philosophy by Glenn Borchardt:

The Ten Assumptions of Science

The Scientific Worldview

Universal Cycle Theory (with Stephen J. Puetz)

Infinite Universe Theory

Published by the Progressive Science Institute

1966 Tice Valley Blvd #172

Walnut Creek, CA 94595

USA

Cover: Birds of a feather in January 1933: Monsignor Georges Lemaître, priest who proposed the Big Bang Theory, and Professor Albert Einstein, inventor of the Untired Light Theory. Credit: Unknown photographer. Taken after Lemaitre's lecture at Mount Wilson Observatory (Lambert, 2020).

Copyright © 2020 by Glenn Borchardt

ASIN: B08N5HYLTB (ebk)

ISBN: 9798559631448 (pbk)(b&w) ASIN: B08N96V91H

ISBN: 9798561794223 (pbk)(color) ASIN: B08NDXBFZP

ISBN: 9798323925131 (hc)(color) ASIN: B0D2R1N74Y

ASIN: B0CY8LBM5R Audible Audiobook

Citation: Borchardt, Glenn, 2020, Religious Roots of Relativity (Version 20241211): Walnut Creek, CA, Progressive Science Institute, 160 p.

Acknowledgements

I thank Marilyn Borchardt, Roger Burbach, Fred Frees, Bill K. Howell, and numerous Blog commenters for many stimulating discussions of the topic. Of course, I am especially indebted to Stephen J. Puetz, my coauthor on "Universal Cycle Theory," the preceding technical volume. Without his outstanding and sagacious input, I doubt we ever would have discovered the physical cause of gravitation. Thanks so much to the following reviewers who provided suggestions that improved the manuscript: Marilyn Borchardt, Bill Howell, George Coyne, Ed Mason, Steve Puetz, Rick Dutkiewicz, Fred Frees, Rudolf Vrnoga, Pierre Berrigan, Mike Gimbel, Steven Bryant, William Westmiller, and Luis Cayetano.

During the last half century, the following institutions generously provided support for my scientific career: University of Wisconsin, Wisconsin Geological & Natural History Survey, Oregon State University, United States Atomic Energy Commission, National Science Foundation, National Academy of Sciences, National Research Council, United States Geological Survey, California Geological Survey, United Nations, and numerous private clients. I wish to thank those who spent endless hours administering these organizations so I would be free to enjoy my explorations in the laboratory and field. I also thank the taxpayers and clients who provided the funds that made these investigations possible. In gratitude, I present this book at minimal cost.

I dedicate this book to:

Marjorie, Arnold, Bertha, Coochie, Roger, Marilyn, Jim, Marion, Francis, Rod, Moyle, Art, Elizabeth, Dennis, Elia, Nina, Hasu, Chuck, Doug, Karl, Harry, Edward, Fred, Bob, Tom, Natalie, Steve, Jesse, and Juan.

Religious Roots of Relativity
Glenn Borchardt

Contents

Acknowledgements 3

List of Tables 8

List of Figures 9

Preface 11

INTRODUCTION 13

CHAPTER 1: IMMATERIALISM 25
The First Assumption of Religion 25

   Immaterialism and Philosophy 26
   Immaterialism and Religion 27
   Materialism and Science 35
   Immaterialism and Relativity 37

CHAPTER 2: ACAUSALITY 49
The Second Assumption of Religion 49

   Acausality and Religion 49
   Causality and Science 52
   Acausality and Relativity 53

CHAPTER 3: CERTAINTY 55
The Third Assumption of Religion 55

Certainty and Religion 56
Uncertainty and Science 57
Certainty and Relativity..57

CHAPTER 4: SEPARABILITY **63**
The Fourth Assumption of Religion **63**

Separability and Religion 63
Inseparability and Science 66
Separability and Relativity 66

CHAPTER 5: CREATION **77**
The Fifth Assumption of Religion **77**

Creation and Religion 77
Conservation and Science 80
Creation and Relativity 81

CHAPTER 6: NONCOMPLEMENTARITY **85**
The Sixth Assumption of Religion **85**

Noncomplementarity and Religion..86
Complementarity and Science 87
Noncomplementarity and Relativity..89

CHAPTER 7: REVERSIBILITY **91**
The Seventh Assumption of Religion **91**

Reversibility and Religion 91
Irreversibility and Science 94
Reversibility and Relativity 94

CHAPTER 8: FINITY **97**
The Eighth Assumption of Religion **97**

Finity and Religion 97
Infinity and Science..97
Finity and Relativity 99

CHAPTER 9: ABSOLUTISM...**103**

The Ninth Assumption of Religion	103
Absolutism and Religion	103
Relativism and Science	106
Absolutism and Relativity	108

CHAPTER 10: DISCONNECTION 111
The Tenth Assumption of Religion 111

Disconnection and Religion	111
Interconnection and Science	113
Disconnection and Relativity	115

CONCLUSIONS 119

References	127
Glossary	133
Appendix	157

List of Tables

Table 1. The Ten Assumptions of Science. 23
Table 2. The Ten Assumptions of Religion. 24
Table 3. Einstein's eight ad hocs. 40
Table 4. Falsifications, contradictions, and paradoxes disproving the Big Bang Theory—a review (modified from Borchardt, 2017).....84

List of Figures

Figure 1. Heaven: What dreams are made of. 29
Figure 2. Hell: What nightmares are made of..30
Figure 3. Interferometer measurements of Earth's velocity around the Sun as determined at various altitudes above mean sea level. These data demonstrate there is an aetherosphere around Earth. Michelson and Morley's sought-for measurement of the "ether wind" was a function of altitude...38
Figure 4. Sid Harris's classic cartoon illustrating scientific skepticism toward an ad hoc. Credit: ©Sidney Harris. 41
Figure 5. Neomechanical interactions apropos the $E=mc^2$ equation illustrating that both absorption and emission involve mechanical collisions described by Newton's Second Law of Motion. By denying aether exists, regressive physics denies that these collisions occur. Credit: Borchardt (2017, figure 25). 71
Figure 6. Pope discourages curiosity. Credit: Jerry Coyne. 79
Figure 7. Dinosaurs in Noah's ark at the Ark Encounter, a Christian theme park funded, in part, by the State of Kentucky. These are supposed to be models of Baryonyx, which became extinct about 125 million years before humans evolved. Credit: Peggy Watson, Citizen of Heaven, Follower of Jesus.......................79
Figure 8. An explosion has only divergence, while creation requires convergence. Credit: Pinterest...86
Figure 9. Cartoonist's take on the entanglement between quantum mechanics and religion. Credit: Mohammed Jones (https://www.jesusandmo.net/)..96
Figure 10. Sid Harris gets it right again: Particles within particles within particles ad infinitum. Credit: ©Sidney Harris.......................99
Figure 11. Some of the infinite number of possible snowflakes. Credit: Snowflake Bentley Website. 107
Figure 12. One reformist's view of regressive physics and cosmogony. . 122

Figure 13. The 30-million increase in the religiously unaffiliated during the last decade in the US. There were 39 million in 2009 and over 68 million in 2019. 123

Figure 14. Religious rejection in relation to youthfulness of generations in the US. Credit: Pew Research Center. 124

Figure 15. Sigmoidal growth curve for global population assuming perfect symmetry about the 1989 Inflection Point (modified from Borchardt, 2007). 125

Glenn Borchardt

Preface

Here is the preface from "Infinite Universe Theory" which was published on December 25, 2017:

"When I first became aware of the Big Bang Theory of the universe, I thought "Wow! That's great; finally, they know how it all began!" Mom, a devout and very conservative Missouri Synod Lutheran, opposed the whole idea of it—her world began with Genesis 6,000 years ago. At billions of years, the timing of the Big Bang was a bit off for her. With an eighth-grade education, she would not be expected to appreciate how fast light travels and how far away those stars and galaxies were. On the other hand, I graduated from a university that favored "fearless sifting and winnowing by which alone the truth can be found." I had even gotten my nose out of the books long enough to look around me. After learning a bit of soil science, I returned home to southern Wisconsin to see the evidence firsthand for an Earth much older than 6,000 years. According to geologists at the university, it turns out I grew up next to a drumlin, an egg-shaped hill that once was under a mile of ice. There were thousands of these drumlins, most with the big end facing directly north and the small end facing south. If you travel east to west in Dodge County, Wisconsin, you will be going up and down drumlins about every mile. Each one, so they said, was over 17,000 years old.

Then I found it. Someone carbon-dated peat from one of the many marshes between the drumlins. The date was 17,700 years. As a literalist born of the Lutheran tradition, I had to make a choice: science or Genesis. I subsequently gathered with my own hands hundreds of samples of peat, wood, and charcoal, with many turning out to be much older than that. Some of the samples dated by other means were millions of years old. During my Postdoc, colleagues dated samples from the Moon at 4.66 billion years old. Guess I chose correctly.

I bring this up because it is similar to the journey I would like you to take in this book. You should begin with a lot of skepticism. After all, what could a farm kid from Wisconsin who plays around in the dirt say about the current theory of the universe? Those smart fellows like Einstein and Hawking surely must have known what they were doing. That is what I thought too—up until 1978.

But that was not to be the case after I started looking into their wild claims in depth. How could the universe explode out of nothing? In my scientific experience, everything came from some other thing. Nothing just popped up out of nowhere. How could the universe be 4-dimensional? Everything I knew had only 3-dimensions. How could the universe be expanding in all directions at once? The claims of curved empty space did not appeal to me. How could light be both a particle and a wave at the same time? In Physics 1a, I was told to leave common sense behind. You needed to know some exceedingly advanced higher math, and besides, only a few physicists could comprehend it anyway. It looked a bit like the emperor's new clothes.

Being skeptical and commonsensical by nature, I investigated "modern physics" systematically from top to bottom. What I found was shocking. The whole thing was an enormous, stinking can of dead worms. The most dubious proclamations were founded on presuppositions that had more in common with religion than science. No wonder modern physics was so popular. I would need to uncover the assumptions that were making physics go awry for over a century. From there I would need to rebuild much of its structure, starting with classical mechanics and incorporating what was missing: the assumption of *infinity*. Happily, not much of this involves complicated math, but if you want to understand this new view of the universe, you need to do some work. In particular, ingrained presuppositions are hard to change. Still, I promise it will be worth it."

I highly recommend your reading Infinite Universe Theory to broaden your understanding of the universe and of the current volume. In this one I focus on the philosophical reasons regressive physics and cosmogony remain so entrenched. It is a rather simple thesis, but one I have not seen anywhere else.

Glenn Borchardt

Walnut Creek, December 11, 2024   Glenn Borchardt

> *Science without religion is lame, religion without science is blind. ...a legitimate conflict between science and religion cannot exist.* -Einstein, 1940[1]
>
> *Science and religion are incompatible.* -Coyne, 2015[2]

INTRODUCTION

The Infinite Universe[3] forces us to make assumptions. Science assumes "all effects have physical causes"; religion assumes "some effects may not have physical causes." I came face-to-face with this contradiction on receiving a third rejection of my paper "The Physical Cause of Gravitation."[4] The editor infamously said "Gravity and its origin are well understood." Note the use of the word "origin" instead of the words "physical cause." Neither Newton's attraction hypothesis nor Einstein's space-time[5] hypothesis specify a physical cause for gravitation. My paper merely suggested since gravitation was an acceleration, then there had to be an accelerator. In other words, a physical cause requires a collider as well as a collidee.

Of course, with gravitation a physical cause is not obvious even though the effects are well known. Such theories are called "kinetic theories." They describe what is happening, but not why it is happening—as far as these theories are concerned, there is no there there. The equations work whether the actual cause exists or whether it involves a miraculous pull or a theoretical push. The mathematics is the same for both in any

---

[1] Einstein, 1940, Science and religion.
[2] Coyne, 2015, Faith Versus Fact: Why Science and Religion Are Incompatible.
[3] Borchardt, 2017, Infinite Universe Theory.
[4] Borchardt, 2018, The physical cause of gravitation.
[5] Linked words are defined in the glossary.

case. Gravitational fields and magnetic fields can just as easily be considered "immaterial" as "material." This is where the relativity-religion connection manifests.

The Dissident Movement

Science and religion are based on opposing fundamental assumptions. But, as I will show in this book, religion and relativity happen to share the same assumptions. This is not true for most scientific theories, which typically are based only upon scientific assumptions. The counter-intuitive claims of relativity have kept it under attack for more than a century. That is because its religious foundation has allowed fabrication of so many weird fantasies surprisingly accepted by employed mainstream physicists as well as the relatively uneducated public. Many of the rest of us have difficulty believing in 4-dimensional "space-time," massless particles having perpetual motion, perfectly empty space, and immaterial fields. Despite perennial claims declaring Einsteinian correctness, unheralded skeptics abound. The Jean de Climont group listed over 9,500[6] people on the web who have objections to relativity, quantum mechanics, and/or cosmogony. The 2,500 proposed alternative theories range from partial reforms involving mathematical tweaks to a few who challenge the very axioms upon which those interlinked disciplines are based.

The complaints are so numerous that mainstream physics has become exceedingly defensive. Any criticism of Einstein or relativity is rejected out of hand.[7] And, as I found out, any claim there must be "physical causes" for the phenomena relativity attempts to explain is likewise rejected as contrary to the sacred paradigm. The upshot is that publication in highly respected journals is next to impossible. As in the various religious cults, opposing views simply are censored with their promoters facing excommunication. True, dissidents can publish in unheralded alternative[8] or infamous predatory journals.[9] But even when these are

---

[6] de Climont, 2020, The Worldwide List of Alternative Theories and Critics. [See Brewis, 2013, for further info on this pseudonymous group.]
[7] Miller, 2019, Column: Sorry, crackpots. [See rebuttal by Bryant, 2019.]
[8] Examples: Physics Essays, General Science Journal.
[9] Dadkhah and Borchardt, 2016, Guidelines for selecting journals that avoid fraudulent practices in scholarly publishing; Dadkhah and others, 2016, Fraud in academic

adequately reviewed, they are seldom cited. Co-authorship among dissidents is almost unheard of. This is because few of them can agree on much of anything. That is why there are so many theoretical attempts to replace relativity and Big Bang Theory.

In this book I will show why the overthrow of relativity is so difficult. In exploring that problem, you will discover its resolution. It starts where every wannabe scientist starts: The Newtonian belief "there are causes for all effects." This belief constitutes the primary scientific faith. It is a "faith" because we never could completely prove that statement to be true for each of the necessarily infinite number of effects. Even so, it has been successful trillions of times, and so we continue to use it. This is not true for the opposing belief "there are no causes for some effects." Neither belief is more "logical" than the other. Each of those beliefs forms the assumptive foundation for subsequent deductions. The first involves the external, real world while the second involves the internal, imagined world. In today's world, both the scientist and the preacher get paid.

Unfortunately, we are unable to find causes for some effects. This leads to interminable debates among philosophers about the universal applicability of the belief in causality. Here we have to get more specific to get rid of much of that doubt. A more complete definition is in order. For that, I will add the word "physical" to the definition. Why is that word important? Is not the word "causes" enough? Apparently not. Superstitious folks are wont to propose "nonphysical" causes for many effects. But the "physical" is what physics is all about. There are no "nonphysical causes."

To understand physics, we have to go back three centuries to Newton's Second Law of Motion, which states all effects are the results of the collision of one thing with another. Object A hits object B, with A slowing down and B speeding up. That's it; that's the gist of physics. We can use an equation to describe the process: $F=ma$, where F is called "force" and "m" refers to the mass of A and "a" refers to how much B's motion was changed as a result of the collision. There is much more to physics than that, of course. Like the Infinite Universe itself, there can be no theoretical end to the detail we might uncover. Nevertheless, ignoring the Second Law leads to a theoretical mess; holding fast to it will make you smarter than Einstein.

In many respects this book is a takeoff on a book I published in 2004.[10] Therein I presented philosophy as the interminable struggle between determinism and indeterminism, which actually are proxies for science and religion. That struggle is progressive because humanity's growth causes it to explore evermore of its surroundings. We are born not knowing the differences between those two ways of looking at the world. As infants, we experiment with the matter all around us, eventually removing that blanket from our heads after it fails to make the universe go away. That is what we will continue to do in this book. I aim to show in detail that the fundamental assumptions underlying relativity are religious rather than scientific.

The War between Science and Religion

In many respects, relativity was a solution to the eternal conflict between science and religion. Numerous devout authors have written about our topic, pointing out relativity's similarities favoring religion. They see this, of course, as a good thing—evidence in support of their supernatural beliefs. Being *natural* scientists, however, we have the opposite point of view. At most, relativity amounts to a temporary lucrative truce in a war that will not end well for either relativity or religion.

Other truces have done much less damage to physics and cosmology. For instance, the late great agnostic, Stephen Jay Gould suggested science and religion are nonoverlapping magisteria (NOMA): They are two different ways of looking at the same thing.[11] In essence, he seems to view

---

[10] Borchardt, 2004, The Ten Assumptions of Science: Toward a New Scientific Worldview.

science as objective and religion as subjective. Science describes objects and predicts what those objects will do; religion proclaims morals and values. Thus, while science can describe events, it cannot pronounce them "good" or "bad" in the name of science—that is solely the bailiwick of subjectivity, religious or otherwise. Religious folks can take heart in that view, with one caveat: They must not contradict the objective conclusions of science in the name of religion, because that is a different magisteria. By the same token science must not contradict the moral and value conclusions of religion in the name of science, because that is a different magisteria. Each of the magisteria are beholden to a different set of teaching authorities, each with a different way of viewing the universe. Would the world be so perfect! Crossing the magisterial line appears more common than not. What religious authorities have not used their positions to encourage followers to disregard legitimate scientific data? What scientific authorities have not used their positions to support morals and values having little to do with their scientific expertise?

For some religions (they seldom agree), the universe was created 6,000 years ago; for some scientists (we don't always agree), the universe was created 13.82 billion years ago. In some quarters Gould's demarcation may have resulted in a temporary truce in the war between science and religion. It also may have given solace to young religious scientists suffering the usual cognitive dissonance they face early in their chosen field. NOMA may work for those with brains that can compartmentalize opposing assumptions. It obviously provides a solution for agnostics who cannot make up their minds or do not want to face the inevitable approbation when they do.

Why We Need to Recognize Fundamental Assumptions

You may ask: What are fundamental assumptions and how did I figure out what they were? Of course, we use nonfundamental assumptions in

---

[11] Gould, 1997, Nonoverlapping magisteria.

science all the time. Often, we call these "hypotheses," statements we test in the external world through observation and experiment. They can be "falsified," that is, they can be proven false even though they never can be proven completely true.[12] Fundamental assumptions are the domain of "metaphysics," that which goes beyond physics. Scientists tend to believe what goes beyond physics simply is more physics, while religious folks tend to believe what goes beyond physics is not physical. According to Collingwood,[13] assumptions are fundamental if they follow these criteria:

1. They have opposites in which, if one is correct, the other is false.
2. They cannot be completely proven or disproven.
3. If one holds two or more assumptions, they must be consupponible, that is, they must not contradict each other.

You also may ask: Why has no other scientist performed this task previously? Please look at Table 1 and notice the 8th Assumption of Science, ***infinity***.[14] The current mainstream assumption is the 8th Assumption of Religion, *finity* (Table 2). In each case, the 8th assumption poses no contradiction with the other assumptions within the appropriate table. The switch from *finity* to ***infinity*** is so radical, however, that it will be the foundation of what I call the "Last Cosmological Revolution."[15] The switch turns physics and cosmology on its head. ***Infinity*** will continue to be resisted with all the might of current physics, cosmogony, and the more than 80% of folks who are religious or previously have used religious assumptions. In science, this "previous use" is particularly important. Like many of us, Einstein himself was confused most of his life. At various times he said:[16]

1923:

---

[12] Popper, 2002, The Logic of Scientific Discovery.
[13] Collingwood, 1940, An Essay on Metaphysics.
[14] The scientific assumptions are in **bold italics** and the religious assumptions are in *italics*.
[15] Borchardt, 2017, Infinite Universe Theory.
[16] Dates not footnoted are from Armstrong, 2017, Albert Einstein's "Cosmic Religion."

> *My comprehension of God comes from the deeply felt conviction of a superior intelligence that reveals itself in the knowable world. In common terms, one can describe it as "pantheistic" (Spinoza).*

1930:

> *I'm not an atheist and I don't think I can call myself a pantheist.*

1940:

> *The situation may be expressed by an image: science without religion is lame, religion without science is blind. ...a legitimate conflict between science and religion cannot exist.*

1941:

> *Then there are the fanatical atheists whose intolerance is the same as that of the religious fanatics, and it springs from the same source . . . They are creatures who can't hear the music of the spheres.*

1949:

> *I have repeatedly said that in my opinion the idea of a personal God is a childlike one. You may call me an agnostic, but I do not share the crusading spirit of the professional atheist whose fervor is mostly due to a painful act of liberation from the fetters of religious indoctrination received in youth.*

1952:

> *There lies the weakness of positivists and professional atheists who are elated because they feel that they have not only successfully rid the world of gods but "bared the miracles". Oddly enough, we must be satisfied to acknowledge the "miracle" without there being any legitimate way for us to approach it. I am forced to add*

> *that just to keep you from thinking that – weakened by age – I have fallen pray to the parsons.*[17]

1954:

> *The word God is for me nothing more than the expression and product of human weakness, the Bible a collection of honorable, but still purely primitive, legends which are nevertheless pretty childish. No interpretation, no matter how subtle, can change this for me. For me the Jewish religion like all other religions is an incarnation of the most childish superstition.*[18]

Einstein's "agnosticism" shows him to be no great student of metaphysics or even of scientific philosophy.[19] The greatest scientists are not so indecisive. They are mostly atheists,[20] with biologists more so than physicists.[21] The disparity between biologists and physicists apparently results from the intensity of the conflict over creationism in each discipline. Biologists struggle against it, while physicists favor it. That fact, and Einstein's fear of "fanatical atheism" should be your first hints as to what is going on between relativity and religion. Those who have thought most deeply about their metaphysics are the most skeptical of religious dogma. One need not be fanatical. One can simply not believe it like the religiously unaffiliated "Nones" do.[22]

In the past, I generally emphasized only the scientific assumptions I discovered, avoiding discussion of their highly controversial religious opposites. This was inadequate. I have since discovered most of the folks

---

[17] Einstein, 1952, Letter to Solovine, March 30.
[18] Einstein, 1954a, Letter to Mr Gutkind impugning religion.
[19] Jammer, 1999, Einstein and Religion: Physics and Theology.
[20] Larson and Witham, 1998, Leading scientists still reject God; Stirrat and Cornwell, 2013, Eminent scientists reject the supernatural.
[21] Gross and Simmons, 2009, The religiosity of American college and university professors.
[22] Pew Research Center, 2019, In U.S., decline of Christianity continues at rapid pace.

trying to reform relativity and cosmology propose modifications founded on religious assumptions. So now, for the first time, I present "The Ten Assumptions of Religion," which you will need for understanding the philosophical foundations of relativity. These are essentially the indeterministic opposites of assumptions I outlined in my book "The Ten Assumptions of Science."[23] This discussion will involve ten chapters, each being an elaboration of one of "The Ten Assumptions of Religion," its antithesis, and its application to relativity. Upon completion, I hope you will have a better understanding about why relativity, quantum mechanics, and Big Bang Theory is still so popular.

---

[23] Borchardt, 2004, The Ten Assumptions of Science.

Glenn Borchardt

Table 1. The Ten Assumptions of Science.[24]

1	Materialism	The external world exists after the observer does not, and that the universe consists of matter.
2	Causality	All effects have an infinite number of material causes.
3	Uncertainty	It is impossible to know everything about anything, but it is often possible to know more about anything.
4	Inseparability	Just as there is no motion without matter, so there is no matter without motion.
5	Conservation	Matter and the motion of matter can be neither created nor destroyed.
6	Complementarity	All things are subject to divergence and convergence from other things.
7	Irreversibility	All processes are irreversible.
8	Infinity	The universe is infinite, both in the microcosmic and macrocosmic directions.
9	Relativism	All things have characteristics that make them similar to all other things as well as characteristics that make them dissimilar to all other things.
10	Interconnection	All things are interconnected, that is, between any two objects exist other objects that transmit matter and motion.[25]

---

[24] Borchardt, 2004, The Ten Assumptions of Science.
[25] [Here we assume, along with Descartes, that "there can't be any empty space in the universe."] Descartes, 1644, Principles of Philosophy, p. 51]

Table 2. The Ten Assumptions of Religion.

1	Immaterialism	Material things have no objective existence, strictly being products of consciousness.
2	Acausality	Some effects have no material causes.
3	Certainty	It is possible to know everything about some things.
4	Separability	Motion can occur without matter and matter can exist without motion.
5	Creation	Matter and motion can be created out of nothing.
6	Noncomplementarity	All things are subject to divergence from all other things.
7	Reversibility	Some processes are reversible.
8	Finity	The universe is finite, both in the microcosmic and macrocosmic directions.
9	Absolutism	Identities exist, that is, any two things may have identical characteristics.
10	Disconnection	There may be perfectly empty space between any two objects.

Glenn Borchardt

CHAPTER 1: IMMATERIALISM

The First Assumption of Religion

Material things have no objective existence, strictly being products of consciousness.

The First Assumption of Religion above seems preposterous or at least dubious to anyone who has accidentally run into any material object. It would seem obvious we are surrounded by things not of our choosing and that those things will be there when we are gone. But we recognize them only through our five senses, which are known to be fallible. Sensory information proceeds to and through our nervous system where it is processed to be imaged as "matter," things that are XYZ portions of the universe and have location with respect to other things. Some objects cannot be sensed easily and others sometimes appear real to us when we dream or otherwise imagine things not real.

The universe also displays a phenomenon we call motion. Motion is not material. Motion is what matter *does*. So, in that sense motion might be confused with *immaterialism*. Support for *immaterialism* may be claimed whenever motion occurs without its material carrier being in evidence. The wind in the willows was evidence for *immaterialism*—until nitrogen and oxygen were discovered. Evidence in support of **materialism** continues to pileup, but because the universe is infinite, a complete proof for that scientific assumption is not possible. The same is true for *immaterialism*. **Materialism** receives support every time we find a material carrier for a particular type of motion; *immaterialism* receives support every time we do not. The upshot is that we must choose between these two fundamental assumptions. Inevitably, there are interminable debates about which to choose, but choosing correctly is of the utmost importance. Otherwise we end up with absurd contradictions and paradoxes typical of relativity and religion.

Immaterialism and Philosophy

Variations on *immaterialism* keep appearing in philosophical and scientific literature. *Immaterialism* requires one to be a bit of a solipsist, a person who tends to be self-centered and not being especially concerned with the outside world. In introductory philosophy, this question often is asked: If a tree falls in the forest, does it make a sound? The materialist knows sound is the vibrations of the air and surroundings, but the immaterialist is not so sure. The tradition enunciated by Bishop Berkeley of "the chair I sat on disappears when I leave the room" fame continues today with Deepak Chopra, a physician who claims we are not "physical machines that have somehow learned to think...[but] thoughts that have learned to create a physical machine."[26] The titles of his books reveal much about where this is going.[27] In particular, it is the weakness of quantum mechanics that it can be used to support the woo-woo that made Chopra rich and famous as a New Age faith healer. There are hundreds of similar books using regressive physics and cosmogony to boost various religious ideas.

Chopra is preying on the solipsism we are all born with. As infants, we can make the external world disappear by putting a blanket over our heads. Humanity's development was no different. At first, we thought we were in a special place created just for us by some benevolent imaginary being. Copernicus straightened that out by showing we were not at the center of all things, but that the Sun was. Even the Sun was nothing special, being

---

[26] Chopra, 2009, Quantum Healing.
[27] Chopra, 2015, How Consciousness Became the Universe; Chopra and Kafatos, 2017, You Are the Universe.

one of the 400 billion stars in the Milky Way galaxy. The fuzzy features on the night sky then became recognized as "island universes"—galaxies much like our own. Today, solipsism takes a hit every time a new galaxy is discovered, with current observations estimating there are over two trillion. And yet, we remain solipsistic, with even highly intelligent cosmologists implying the entire universe was created out of nothing or of their imagined singularity.

Immaterialism and Religion

Religion *could not exist without immaterialism*. One obvious reason religion needs *immaterialism*, is the fact it is impossible to provide material support for its claims. No one has discovered the physical location of "heaven" or "hell" or of the hypothesized creator of the universe.[28] That is because those things do not and cannot possibly exist. They simply are the dreams and imaginings of the solipsist.

Dreams and the Origin of Religion

Dreams, which are necessary motions within the brain, have been considered evidence for *immaterialism* ever since animals developed enough consciousness to dream. According to David Zeigler, religion itself originated in dreams:[29]

> *To start with, dreams seem to support a belief in the dualism of mind and body. When we dream, we experience events and places that the body does not, thus it seems there is a separate reality for the mind or spirit.*

---

[28] Hooper, 1903, Aether and Gravitation. [In this otherwise informative book, the author finishes by declaring the creator of the universe was located in the center of our Milky Way galaxy.]

[29] Zeigler, 2020, Religious belief from dreams?

Without a belief in mind/body dualism, it is hard to imagine any reason for religious belief to have ignited.

Certainly, our ancestors of prehistory recognized one another as individuals and likely dreamed about those recently dead members of their tribe or group. This experience (along with the idea that spirits of the living could leave their bodies during dreams) could easily have given rise to belief in a spirit world where the dead ancestors lived and where the living could make visits during dreams.[30] Such a belief would mean that these dead ancestors are still around and still interacting with the living, perhaps trying to communicate with and guide us. In short, ancestor worship could easily spring from these assumptions, and ancestor worship is an old and widely spread belief in many of the world's primitive religions.[31]

...the Bible contains several instances of God speaking to or instructing individuals in dreams. In one such example in the book of Matthew, Joseph is instructed in a dream to marry Mary even though she is pregnant because the child is "of God" and therefore no shame should be felt by him or Mary.

Some folks take their dreams to the external world, proclaiming a new religion, talking to grave stones, or celebrating the dead. The latter was always a puzzle to me when I heard of the "Day of the Dead," a public holiday in Mexico, which originated with the Aztecs 3,000 years ago. Although first opposed as sacrilegious, it now has been incorporated into the Catholic religion. People tend to dream about the things and events

---

[30] Howells, 1949, The Heathens.
[31] Ibid.

common to their environments (Figure 1 and Figure 2). That is why ancient religions include shepherds and sheep and modern religions include aliens and science fiction. There are so many disparate religions because there are so many disparate dreams, each produced from a unique point of view. In a way, the segmentation of so many dreams into *only* 4,200[32] religions is quite remarkable. It attests to the ability of humanity to organize societies in carrying out the main evolutionary purpose of religion: to instill and enforce loyalty. Throughout history, tribes, states, and countries with a loyal following were better able to succeed in conflicts over the resources needed for survival.

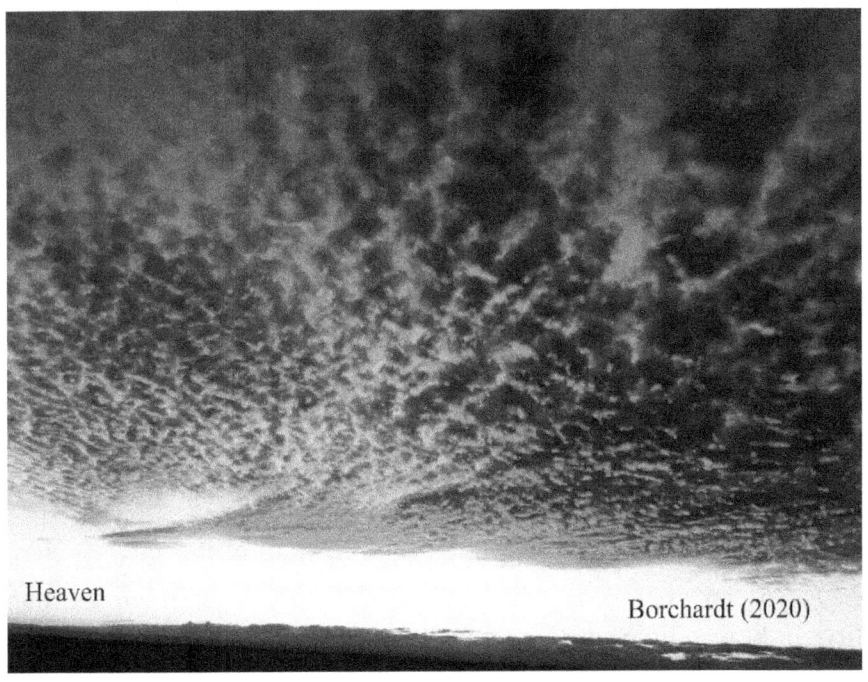

Figure 1. Heaven: What dreams are made of.[33]

---

[32] https://go.glennborchardt.com/4200-religions
[33] Sunset from the Sacramento Valley on November 11, 2017. Credit: Glenn Borchardt.

Figure 2. Hell: What nightmares are made of.[34]

Distinguishing between dreams and reality is often difficult.[35] When I hear a knock on the door in my dream, I am not sure if it is real or not. Sometimes it is incorporated into the dream, and at other times it wakes me up to the reality that someone actually is knocking on the door. Making the distinction between ideality and reality is the crux of the problem. If your material life is not so great, you might prefer the ideality of your dreams. The "American dream" is of an imagined future better than the current reality. In both religion and science, we always must be on guard against

---

[34] Mouth of the Halema'uma'u Crater of Hawaii's Kilauea Volcano in 2014. Credit: Andrew Hara/USGS.
[35] Rassin and others, 2001, When dreams become a royal road to confusion.

"confirmation bias," in which we subconsciously overemphasize data that supports our dreamt-up theories and neglect the data that does not.

Dreams are as important in science as they are in religion and everyday life. The builder must imagine the structure before the drawings can be made and the actual construction can begin; the scientist must imagine the chemical structure before the experiments can be designed and performed. This is, after all, where ideas come from. As professor M.L. Jackson once told me: "I read chemistry and mineralogy papers during the day and come up with new ideas by combining the information in my dreams at night."

There is a famous story about the structure of benzene, which was discovered by August Kukulé in 1862. It seems he was sitting by the fire, day-dreaming about benzene's 6-carbon chain. In the dream, several carbon chains were dancing around, when one of the chains bit its tail, forming a ring. Voila! The benzene ring is probably one of the most famous structures in all of organic chemistry.

The 4,200 dissimilar religions in the world demonstrate daily that dreaming can get out of hand. Regressive physics and the 2,500 theories opposing it are similar. The only significant difference between the religious and the scientific dreams is the requirement that the scientific ones be supported by observations and experiments performed in the external world. Even then, mistakes can be made, as they are in regressive physics when incorrect fundamental assumptions are used to interpret data. In the current context, the best example is Einstein's use of his "Untired Light Theory" and his "4-dimensional space-time" to support the expanding universe nonsense.

Attempts to Justify Immaterialism

While surrounded by obviously material objects, theologians (and some religious scientists) go to great extents to justify *immaterialism*. Here is one such, rather forthright view about the challenge provided by **materialism**:

...certain philosophical doctrines are incompatible with religion. Thus, for example, unless it is possible to have objective knowledge of non-sensible things, belief in God makes no sense.[36]

This is quite an admission by a religious fellow. The only way we can have knowledge of objects is to sense them. No one could have knowledge of "non-sensible things." Such so-called "objective knowledge" certainly has nothing to do with "objects." It is thus not "objective," but purely imaginary.

In my experience, most preachers do not seem to know what "materialism" is. The word itself tends to be used more often by its other definition as a proxy for the rampant over consumptive characteristic of our age. The more learned theologians do occasionally address **materialism**:

Chris [Stefanick] noted the dispute identified as "science vs. religion" is more accurately called "materialism vs. religion." Materialism is the belief that the only place we can find truth is by looking at the physical world. According to a materialistic worldview, God must not exist because we can't see or touch God. Stefanick gave several examples of how reason can lead us to believe in God's existence, asserting that believing the universe occurred by random chance, without the action of a creator, is like believing that an explosion at a print shop could produce an unabridged dictionary.[37]

Of course, theologians like Chris Stefanick would not be able to get it. I put his first-cause argument to bed years before in "The Scientific Worldview,"[38] in which I explained that the universal mechanism of

---

[36] Walbridge, 2001, The Most Learned of the Shia, p. 134.
[37] https://www.thewitnessonline.org/catechesis/chris-stefanick-reboots-marion-parish/

evolution is univironmental determinism (what happens to a portion of the universe is dependent on the infinite matter in motion within and without).[39] An infinite universe does not need a cause. There always is yet another thing to do the colliding. Being infinite, it turns out the universe appears utterly pragmatic. Every motion of each thing, from the pairing of two disparate aether particles to the formation of the 100-billion neuron brain to the formation of the largest galaxy cluster is a result of the "destruction of the unfittest." No portion of the universe can survive an antagonistic univironment. Obviously, I sympathize with Stefanick's doubt that anything can form via things coming apart, with the claims of the Big Bang Theory being archetypically the most absurd. On the other hand, the randomness he uses as a crutch does not occur either, as we will see in the chapter on *certainty*.

Creationists, who are invariably religious and opposed to neo-Darwinian evolution, tend to be adamantly against **materialism**. In the description of his book, Professor Stephen Barr, not surprisingly a regressive physicist, tried to make the over-worn case for science without **materialism**:

> *A considerable amount of public debate and media print has been devoted to the "war between science and religion." In his accessible and eminently readable new book, Stephen M. Barr demonstrates that what is really at war with religion is not science itself, but a philosophy called scientific materialism. Modern Physics and Ancient Faith argues that the great discoveries of modern physics are more compatible with the central teachings of Christianity and Judaism about God, the cosmos, and the human soul than with the atheistic viewpoint of scientific materialism.*[40]

---

[38] Borchardt, 1984, 2007, The Scientific Worldview.
[39] Borchardt, 2007, The Scientific Worldview: Beyond Newton and Einstein.
[40] Barr, 2003, Modern Physics and Ancient Faith.

Barr, and others like him, will come up throughout this discussion. Much of what I am trying to do—show that relativity and religion use the same indeterministic assumptions—has already been done by physicists like Barr and Rabinowitz[41] who use the false assumptions of regressive physics to support religion. The task for you as a reader is to decide whether that situation is better for physics or for religion. That decision was easy for Professor Mark Perakh who reviewed Barr's tome and concluded:

> *Without having laid the foundation for his conclusion, Barr asserts that the human mind results from a special creative action of some unobservable entity referred to as God, but proclamations of faith are hardly persuasive for skeptics.*
>
> *Barr is entitled to his beliefs, but in my view his assertions of the fallaciousness of "materialism" have nary a chance to change the mind of a single skeptic.*[42]

In an interesting compromise with **materialism**, one priest came up with the idea of "Christian materialism":

> *Authentic Christianity, which professes the resurrection of all flesh, has always quite logically opposed 'disincarnation,' without fear of being judged materialistic. We can, therefore, rightfully speak of a Christian materialism, which is boldly opposed to those materialisms which are blind to the spirit.*[43]

This highlights one of the primary confusions suffered by immaterialists: Mistaking motion for spirit. Because the Infinite Universe only has two basic phenomena, to be "blind to the spirit" would be to assume matter

---

[41] Rabinowitz, 2006 Mindless materialists.
[42] Perakh, 2007, Non sequitur in five parts: Professor Barr's effort to bolster his faith via modern physics and Gödel's theorem.
[43] Pazukhin, 1997, The Christian materialism of Blessed Josemaría Escrivá.

without motion—a violation of *inseparability*. We all are "blind to the spirit" if motion = spirit, for we cannot see motion. We can only "see" the thing that moves.

Materialism and Science

Science could not exist without materialism. One obvious reason science needs *materialism*, is the fact it is impossible to do science without observing or experimenting with matter. True, one can think about matter without observing or experimenting with it, but that alone, is not science. An inperiment ("thought experiment") is not a true experiment, because all *ex*periments must deal with the *ex*ternal world outside ourselves. While religion discourages such curiosity; science reveres it. Contrary to Berkeley and Chopra, true scientists believe the universe *ex*ists independently of consciousness. That is why we simply define the First Assumption of Science, *materialism*, as the claim "the external world exists after the observer does not."

Matter and Mass

Next, we need to be more specific about what this external world amounts to. For that we must define matter as stated in the Glossary:

> MATTER. *An abstraction for all things in* existence. *Above all, matter always contains other things within and without, ad infinitum. There are two basic types of matter: baryonic and aether. Although baryonic matter is what we ordinarily observe, aether is tiny and normally not directly detectable. Both have mass produced by constituents subject to interactions demonstrated by Maxwell's* $E=mc^2$ *equation. The "solid matter" of the idealist does not exist.*

Mass, although dependent on matter, has an operational definition:

> MASS. *Resistance to acceleration. We determine the mass of a* microcosm *by accelerating it with another*

microcosm of known mass and velocity. Because all microcosms in the universe are in motion, mass and velocity are relative, so we have established standards, which on Earth, are relative to the acceleration of gravity (9.81 m/s²). Because length and time *also are relative, we have established conventions for those too. Realize, however, that these conventions are not absolute. Because the universe is infinite, measurements for each of them have a plus or minus and each tends to change over time and location. That is why we occasionally add a leap second to the length of the day as Earth's rotation rate slows. Although related, mass and matter are not identical. Mass is dependent on the internal motion of* submicrocosms, *which increases with the absorption of motion and decreases with the emission of motion per Maxwell's $E=mc^2$ equation.*[44]

Every portion of the infinite universe contains matter within and without. Nonexistence (the idealist's perfectly empty space) is impossible. Both solid matter and empty space are imaginary endpoints of the matter-space continuum. All real objects appear to have both characteristics.

Before the regression in physics **materialism** underwent strong development. In 1859 Darwin and Wallace proposed materialist evolution in biology; in the 1880s Marx proposed dialectical and historical materialism; and in 1883 Engels was one of the first to suggest life originated from inanimate matter. In 1908 Lenin published "Materialism and Empirio-criticism" a polemic against the reaction to **materialism** in science that was gaining strength even before Einstein became an icon.

---

[44] Borchardt, 2009, The physical meaning of $E=mc^2$; Maxwell, 1877, Matter and Motion.

The use of materialism as the proclaimed philosophy for the overthrow of capitalism in Russia, China, and elsewhere made the word non grata in the US during the Cold War. Religion was promoted in opposition, with the government endorsing it on coins, currency, and in the pledge of allegiance in the 1950's. In the US, to be a nonbeliever was considered as nothing short of treasonous. Scientists, of course, continued to be materialists at least during their working lives, although they often dared not acknowledge it.

Immaterialism and Relativity

Science is based on **materialism**, but Einstein's *immaterialism* and associated solipsism exists throughout relativity, which therefore is ipso facto not scientific. This all started with his radical rejection of the theoretically necessary aether—setting into motion the counter-revolution in theoretical physics I call a regression. In its place he substituted perfectly empty space—the immaterialist's dream.

Empty Space

Unrecognized by most regressive physicists is the fact that the nearly null results of the Michelson-Morley experiment[45] occurred because the sought-after aether formed an "aetherosphere" around Earth similar to the atmosphere. In essence, their experiment was like trying to measure the jet stream in your backyard. Both the atmosphere and decelerated aether are attached to Earth. Subsequent measurements, dismissed by regressives, were a function of altitude (Figure 3). The sought-for 30 km/s "ether wind" could not be measured at altitudes lower than 500,000 m.

---

[45] Michelson and Morley, 1887, On the relative motion of the earth and the luminiferous aether.

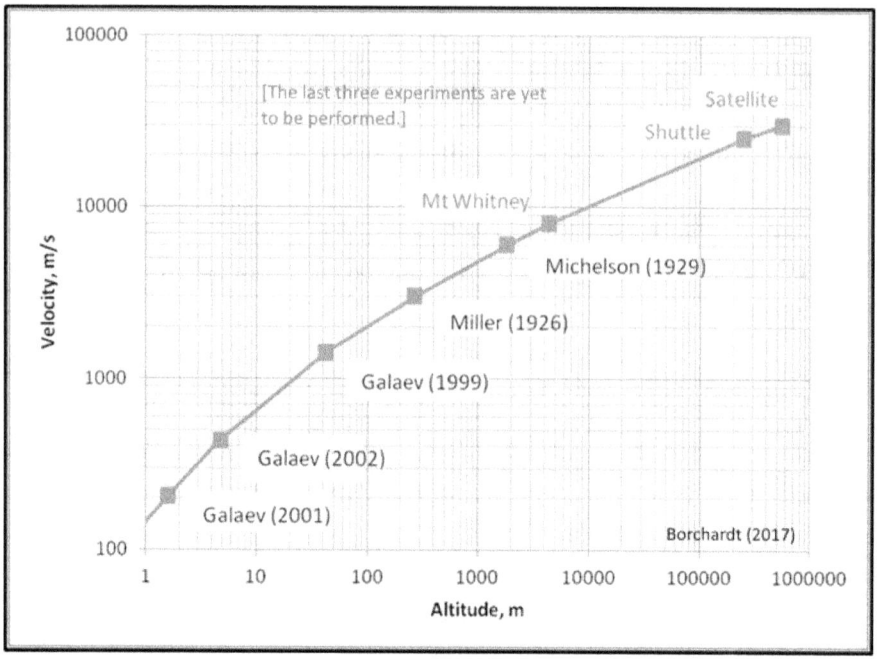

Figure 3. Interferometer measurements of Earth's velocity around the Sun as determined at various altitudes above mean sea level. These data demonstrate there is an aetherosphere around Earth. Michelson and Morley's sought-for measurement of the "ether wind" was a function of altitude.

With the elimination of ether, Einstein had a special problem that he solved with still more *immaterialism*. His only Nobel was for the photoelectric effect. That occurs when light impacts an electrode, producing an electrical current. That was the start of his particle theory of light, with this imaginary particle subsequently being called a "photon" by Lewis in 1926.[46] The photo-electric effect indeed was caused by particles, but they were simply local aether particles that comprised the aether medium disturbed by light waves having the proper frequency. Unlike the imagined photons, these aether particles only had short-range travel—

---

[46] Lewis, 1926, The conservation of photons.

similar to the nitrogen molecules that make up the air through which sound waves travel. The imagined photons are supposed to travel from galaxy to eyeball; aether particles do not. In experiments, many of the properties attributed to massless photons actually are those of aether particles having mass. Massless particles, even if they miraculously existed, could not cause anything. This is because causes are produced by collisions defined by the equation F=ma. Thus, if m=0, then F would be 0.

Einstein's counter-revolutionary claim that light was a particle traveling through space devoid of aether was tested by Georges Sagnac eight years thereafter.[47] His experiment was exceedingly clever, but quite complicated.[48] In brief, he showed light was not a classical particle and that the fringes he got in his rotating interferometer were best explained by the presence of the aether medium. That was the scientific interpretation.

Einstein's religious interpretation required eight ad hocs (Table 3). Although usually frowned upon by scientists, ad hocs are augmentations used to prevent a theory from being disproven (Figure 4). Still, it is unusual to have so many. Nevertheless, that is what it takes to consider wave motion in a medium to be particle motion in empty space. I doubt if you will find anything like Table 3 in the literature. Anyone favoring the photon theory at least has to accept the ad hocs subconsciously. Regressives and reformists generally do not know what to make of the Sagnac experiment. Like the rest of relativity and religion, debates about it are interminable. Einstein himself was relieved to receive support for his idea light was not a classical particle, but a special particle of which he was the sole inventor. Had Sagnac shown light to be a classical particle, Einstein's mathematical circumlocutions would have been for naught. Sometimes it is better to be lucky than smart.

---

[47] Sagnac, 1913a, The demonstration of the luminiferous aether; Sagnac, 1913b, On the proof of the reality of the luminiferous aether.
[48] Borchardt, 2017, Infinite Universe Theory, Chapter 15.1. [This is a detailed explanation of Sagnac.]

Table 3. Einstein's eight **ad hocs**.

1	Unlike other particles, Einstein's light particle always traveled at the same velocity—it never slowed down.
2	Unlike other particles, it attained this velocity instantaneously when emitted from a source.
3	Unlike other particles, it would not take on the velocity of its source.
4	Unlike other particles, it was massless.
5	Unlike other particles, light particles did not lose motion when they collided with other light particles.
6	Unlike other particles, any measurement indicating light speed was not constant had to be attributed to "time dilation"—another especially egregious ad hoc.
7	Time had to be considered something other than motion, for motion cannot dilate.
8	He had to ignore the fact that, unlike particles, the velocities for wave motion are constant because they are dependent on the properties of the medium.

Figure 4. Sid Harris's classic cartoon illustrating scientific skepticism toward an **ad hoc**. Credit: ©Sidney Harris.

Let us now explore the ad hocs directly founded on *immaterialism*. It is ironic that the 1st ad hoc (Unlike other particles, Einstein's light particle always traveled at the same velocity—it never slowed down) would have been rejected by the young Einstein when he was working at the patent office in Bern. Perpetual motion of a single particle would be a violation of the Second Law of Thermodynamics. In this case, however, the usual excuse is the corresponding assumption that space is perfectly empty and there would be nothing to slow it down—*immaterialism* at its finest, although empty space never has been found anywhere. This gets immediately to the heart of what Einstein was trying to do: account for the relative constancy of light. Unlike particles, the velocity of waves through a medium is controlled by that medium—it actually *is* relatively constant. For example, bullets slow down as they travel through air, but the sound waves produced by the weapon do not. Thus, the correct interpretation is that light has a relatively constant velocity merely because it is a wave, not a particle.

The 4th ad hoc (Unlike other particles, it was massless) likewise is the progeny of *immaterialism*. According to one of the most prominent regressive physicists, C.W. Misner:

> *Einstein showed us that immaterial entities are fundamental constituents of the universe. ... But beyond Einstein and modern physics we find many other examples in modern culture of the expanding conquest of immaterial entities while material objects decrease relatively in value, although apparently overwhelming us.*[50]

Nothing has changed during the four decades since the Misner quote. Regressive physicists, even if atheists, still defend the absurdity:

> *In conclusion, a photon passes all tests of materiality. The soul, universal consciousness, qi energy, etc., fail*

---

[50]Misner, 1978, The immaterial constituents of physical objects.

them; they are simply figments of the imagination of generations of mystics and theologues. No commonality exists between massless objects and mystical concepts.[51]

This assertion is absurd. All things in the Infinite Universe consist of other things ad infinitum, and all exhibit the property of mass (i.e., resistance to acceleration). One cannot even imagine what kind of beast could be filled with nothing at all and still travel at 300 million meters per second through nothing at all. Light, as mentioned, is the motion of matter and motion does not have mass. That is why Einstein's imaginary particle had to be massless. Another reason is the math Einstein used in relativity. Dividing its "rest" mass by the Lorentz Correction Factor would result in an infinite mass if the photon had any mass at all. Even Einstein knew that would not make sense.

Einstein's 6th ad hoc: "Time Dilation"

This brings us to Einstein's 6th ad hoc: "Unlike other particles, any measurement indicating light speed was not constant had to be attributed to "time dilation"—another especially egregious ad hoc." (Table 3.) Time is motion, something Einstein never understood. Otherwise, he never would have devised relativity. Only things can dilate or contract; their motions cannot. His objectification or reification of motion in his misinterpretation of the Lorentz Correction Factor in SRT and use of his imaginary "4th dimension" in GRT was his greatest philosophical error.[52] I can't imagine what regressives are thinking, but this error is still prominent in theoretical physics. Ordinary folks scratch their heads when faced with the inanities of relativity. Being raised on a Wisconsin dairy farm, I couldn't get it either. One of the only courses I got a "C" I in was first-semester physics, which was taught by a firm believer who used a trendy textbook that began with a

---

[51] Hassani, 2019, Massless is not nonmaterial. [My rebuttal is at: https://go.glennborchardt.com/Hassani].
[52] Borchardt, 2011, Einstein's most important philosophical error.

cartoon version of relativity. Fortunately, the second semester was by a superb 83-year old professor who taught mostly Newtonian mechanics from an excellent textbook.[53] I got an A. Surprisingly, one of the most popular posts on our PSI Blog is entitled "Time is Motion." Readers often comment they suspected as much for many years and are gratified to see my explanation at the top of the Google search list.

Relativity of Simultaneity

Einstein was a positivistic solipsist. That is almost the definition of today's theoretical physicist. It is an affliction prone to those who do not get out of the lab enough. One puzzlement was his insistence SRT was necessary to understand simultaneity:

> *According to Einstein's special theory of relativity, it is impossible to say in an absolute sense that two distinct events occur at the same time if those events are separated in space.*[54]

Actually, the truth is that you can say that "two distinct events occur at the same time," but you can never prove it. You could only *assume* it because any proof requires communication, which cannot occur faster than the speed of light. Thus, if light leaves the Moon at the same time as light leaves the Sun, the light from the Moon will arrive on Earth in about 1.3 seconds and the light from the Sun will arrive on Earth in about 500 seconds. Those values can be calculated simply by knowing the distance to the Moon (384,400 km) and Sun (150 million km). You also have to take into account your own motion and its direction. That is a no-brainer, a simple calculation having nothing to do with relativity. It *does* rely on the fact everything in the universe is in motion and the fact wave motion through the aether occurs at c.

---

[53] Sears and Zemansky, 1960, College Physics.
[54] Wikipedia.

Now for the positivism. These calculations rightly demonstrate it is impossible to prove both the Moon and Sun exist at the same time. For instance, if the Sun suddenly disappeared, you would not know about it for 8 minutes and 20 seconds. We agree with Einstein that the *measurement* of simultaneity depends on the location or "frame of reference" of the person doing the measuring. If you were 388,400 km from the Sun, the above measurements would be reversed. Some immaterialists have considered this as proof against "absolute simultaneity," the realistic, common-sense view we all hold that any two things in the universe can exist at exactly the same time. You rightly assume your friend in New York exists at the same time you do. In other words, unless you were Bishop Berkeley, you would assume the chair you are sitting in and the chair in the other room exist at the same time.

Since Einstein, however, the subject of "absolute simultaneity" has been part of an interminable debate among regressives and reformists. For instance, one such has come up with an answer that does not abandon relativity altogether: "…the simultaneity of distant events can be a matter of convention only in a four-dimensional world."[55] Egads! The arguments continue, trying to resolve the paradox between relative and absolute simultaneity. Of course, as I have maintained for decades, paradoxes exist merely because they are founded on one or more false assumptions.[56] In general, like other empiricists, positivists do not admit to holding fundamental assumptions that by definition cannot be completely proven.

The paradox is resolved by a steadfast belief in ***materialism***. All matter exists at all times despite what we think about it. We also assume, like Einstein implied, that all matter is in motion at all times. This means that although absolute simultaneity cannot be measured in real time, it

---

[55] Petkov, 1989, Simultaneity, conventionality and existence. [All he is saying is that time (motion) must be taken into account during measurements designed to support absolute simultaneity. His implication that the world is four-dimensional is merely superfluous, obsequious flatter of Einstein.]
[56] Borchardt, 1984, 2007, The Scientific Worldview.

needs to be assumed. Any measurements anyone could make about the locations of things in the universe always will be in arrears. Even that chair in the next room may have been removed by someone when we were not looking. So what? That does not stop us from assuming that chair or its remains still exist somewhere in the universe.

Immaterialism and Fields

As mentioned above, Einstein was an agnostic. This ambivalence was crucial in his formulation of relativity. At the time, there was no known causative agent for gravitation or magnetism. It was anyone's guess as to whether it was material or immaterial. Einstein equivocated about this. At first, he accepted Michelson and Morley's misinterpretation that aether did not exist, forthwith removing any possibility of discovering the material, causative agent for gravitation and magnetism as well as the medium for light.

In 1905 he wrote:

> *The introduction of a 'luminiferous ether' will prove to be superfluous inasmuch as the view here to be developed will not require an 'absolute stationary space' provided with special properties.*[57]

Here Einstein was properly rejecting the fixed ether apparently disproven by the MMX experiment. Unfortunately, his theory threw out the baby (aether) with the bathwater. That was strange for someone who got famous for insisting everything in the universe was in motion. Why not aether? And why wouldn't aether particles behave according to Newton's laws of motion just like all the other bodies in the universe? Aether was anything but superfluous, as shown by his having to devise eight ad hocs just to consider light as a massless particle traveling through perfectly empty space. Unfortunately, this is where regressive physics is now stuck, with

---

[57] Einstein, 1905, On the electrodynamics of moving bodies, p. 38.

aether denial being its paramount petard. Without aether, there can be no material body responsible for the acceleration that causes gravitation. Einstein himself was not so sure, although much of his musing was confusing.

In 1920 he said:

> *Careful reflection teaches us that special relativity does not compel us to deny ether. We may assume its existence but not ascribe a definite state of motion to it ..."* "*There is a weighty reason in favour of ether. To deny ether is to ultimately assume that empty space has no physical qualities whatever.*[58]

In 1950 he wrote:

> *According to general relativity, the concept of space detached from any physical content does not exist. The physical reality of space is represented by a field whose components are continuous functions of four independent variables-the coordinates of space and time. It is just this particular kind of dependence that expresses the spatial character of physical reality. Since the theory of general relativity implies the representation of physical reality by a continuous field, the concept of particles or material points cannot play a fundamental part, nor can the concept of motion. The particle can only appear as a limited region in space in which the field strength or the energy density are particularly high.*[59]

In 1954 he wrote:

---

[58] Einstein, 1920, Ether and the theory of relativity.
[59] Einstein, 1950, On the generalized theory of gravitation.

I consider it quite possible that physics cannot be based on the field concept, i.e., on continuous structures. In that case, nothing remains of my entire castle in the air, gravitation theory included, [and of] the rest of modern physics.[60]

In 1980 an appropriately named disciple wrote:

Thus we arrive at the notion of a potential field which is a real physical presence, although an immaterial one in the sense of its not consisting of atoms or molecules of matter.[61]

Regressive physics still denies aether and provides us no clue as to what exists within these fields that should cause the wonderful actions of gravitation and magnetism. Without a physical cause, these phenomena currently appear as magic, in tune with the religious view opposing **materialism**.

Einstein was correct that fields cannot be "continuous structures." All effects, all events occur because of collisions between microcosms, so gravitational and magnetic fields must contain particles that do that job, with aether particles being the most likely candidates. Without aether, fields may as well be considered "immaterial." Newton wavered similarly, offering two possibilities for gravitation: 1) a magical immaterial pull obeying none of his three laws of motion at one time and 2) a material push obeying his Second Law of Motion at another. His famous equation was kinetic. It did not require a physical cause and could be used to describe the effect whether it was a pull or push.

Now let's go to the Second Assumption of Religion, *acausality*.

---

[60] Einstein, 1954b, Ideas and Opinions.
[61] Angel, 1980, Relativity: The Theory and its Philosophy, p. 178.

Glenn Borchardt

CHAPTER 2: ACAUSALITY

The Second Assumption of Religion

Some effects have no material causes.

As mentioned, the universe only displays two phenomena: matter and the motion of matter. The first assumption of science, **materialism**, posits the existence of matter; the first assumption of religion, *immaterialism*, denies that. The second assumption of science, **causality**, posits that, instead of being fixed in space, these material things are in motion, interacting with each other to create the effects we see all around us. The second assumption of religion, *acausality*, either denies this entirely, as in the case of Chopra, or assumes certain effects receive help from an imagined immaterial supernatural power.

Acausality and Religion

The causes of effects are not always obvious. When large object A collides with large object B, it is clear what caused the acceleration of object B. But what happens when the colliding object is so tiny it cannot be seen or detected? This characteristic of the Infinite Universe leaves much room for speculation by religious folks. Most would not agree the universe only consists of material objects colliding with other material objects. Surely, they might say: "There must be something more than that, something not physical that produces many of the events we see all around us." You can hear such complaints during most any Sunday sermon.

Religious "scientist" Dr. Rupert Sheldrake voiced his opposition to strict physicality, even implying it would lead to the downfall of science itself:

For more than 200 years, materialists have promised that science will eventually explain everything in terms of physics and chemistry. Believers are sustained by the faith that scientific discoveries will justify their beliefs. The philosopher of science, Karl Popper, called this stance 'promissory materialism' because it depends on issuing promissory notes for discoveries not yet made. Despite all the achievements of science and technology, materialism is now facing a credibility crunch that was unimaginable in the twentieth century.[62]

Let's see if there really is a "Sheldrake credibility crunch." In their struggle with religion at the monetary table some scientists adopted the belief that "if I cannot see, feel, hear, taste, or smell it, it does not exist." This was convenient for disavowing claims for the existence of spirits, ghosts, gods, souls, and other "super" natural things. It got carried away, however, when that philosophy (positivism) was applied more generally. As mentioned previously, in formal, mathematical physics, theories in which no material cause is evident (e.g., gravitation and magnetism) are called "kinetic." That is, an acceleration (i.e., effect) may be observed, but the perpetrator of that acceleration may not be obvious. As with Newton's equation for gravitation, the math works even though a physical cause is not obvious.

Carl Jung got famous for promoting the idea of *acausality* via his failure to find a cause:

How are we to recognize acausal combinations of events, since it is obviously impossible to examine all chance

---

[62]Sheldrake, 2012, The Science Delusion. [Sheldrake is a former biochemist who, like Deepak Chopra, is well known for his pseudoscientific endeavors, particularly in parapsychology.]

happenings for their causality? The answer to this is that acausal events may be expected most readily where, on closer reflection, a causal connection appears to be inconceivable.[63]

Like Sheldrake, and other positivists, Jung tried to use this human "failure to explain" to support his quasi-religious belief in paranormal phenomena. But as is often stated by nonpositivistic scientists, "the absence of evidence is not evidence of absence." This is stated formally as the Second Assumption of Science, *causality* (All effects have an infinite number of material causes) (Table 1). In other words, in a universe where matter is infinitely subdividable, it is never possible to determine all the causes for any effect, because they are infinite. That is why every measurement has a plus or minus. Just because we cannot detect or conceive of all the causes involved in one of Jung's coincidences does not mean they don't exist. He wasted his time with the "study" of extrasensory perception (ESP) merely because he did not understand perception. All detections require a sensing device capable of receiving microcosmic collisions from the environment. If there really was a "sixth sense," it too would have to involve a sensor. It then would not be "extrasensory." Thus, ESP cannot be real and has no scientific basis. Though popular with religiously indoctrinated folks and their New Age offspring, such claims have been debunked endlessly.[64]

More recently, we have this from Dr. Avi Rabinowitz, a rabbi/physicist:

Big bang theory proposes that the universe's age is finite, and so one can imagine the universe as having somehow emerged into being; we consider such an emergence to be 'acausal'. Given that the universe exists due to acausality, we propose that such acausality is actually a fundamental feature in the universe's operation. Specifically, it can be

---

[63] Jung, 1960, Synchronicity: An Acausal Connecting Principle.
[64] This is a specialty of a magazine called Skeptical Inquirer.

seen as a resolution of quantum physics' measurement problem, and as providing an essential ingredient for 'true free will'.[65]

Here, the connections between the assumptive foundation of relativity and religion is obvious: theologian-physicist-acausality-BBT-cosmogony-creation-free will. As I explained in "The Scientific Worldview," the endless, futile debates about free will basically are debates between scientifically trained folks who assume causality and religiously trained folks who assume acausality. The debates are interminable because causality and acausality are fundamental opposites—if one is correct, the other is incorrect. Like the god/no god debate, a complete proof for either of them is impossible, Because the universe is infinite, we never will be able to determine all the causes for even one effect. I repeat, all fundamental assumptions by definition, have opposites that are either correct or incorrect. To advance knowledge we simply must choose the correct assumptions; to retard knowledge we must continue to choose the incorrect assumptions.

Causality and Science

Causality

It is important to know what **causality** is. As alluded to above, it simply amounts to the collision of one thing with another. Newton's version is what I call "finite universal causality."[66] I call the new, scientific version of **causality** "infinite universal causality," which is stated as: "All effects have an infinite number of material causes." This is my modification of Newton's Second Law of Motion. Note that, like the other assumptions of science, this is consupponible with **infinity**. Newton's law

---

[65] Rabinowitz, 2006 Acausality: The root of "true free will" & of universal emergence into existence.
[66] Borchardt, 2007, The Scientific Worldview.

was a first approximation: one thing in imagined empty space colliding with another in imagined empty space. But as mentioned above, perfectly empty space does not exist—it is only an idea. That means the Second Law never can occur exactly as portrayed by Newton. Per ***infinity***, there always is yet another thing interfering with any particular collision. However, as practical scientists, we try to reduce these interferences to a minimum by "controlling the experiment." At some point, however, we will be inclined to give up, finally tabulating a few of the causes for a particular effect. Often, that is "good enough for government work," as we used to say. But Newtonian mechanics got branded with what I call "finite causality," that is, the assumption there actually could be a finite number of causes for effects. That is what "Laplace's Demon" was all about. With a finite number of causes for each effect, this imaginary being could predict the future and postdict the past with perfect precision.

The Infinite Universe would not allow that, of course. Every determination always has some plus or minus. Everything in the universe is bathed in a macrocosm containing trillions of "supermicrocosms," that is, other objects, molecules, atoms, and aether particles. Laplace's required perfect precision was just as imaginary as the perfectly empty space required for its enactment. Sheldrake's complaint against ***materialism*** amounts to a complaint against finite causality and the imagined *certainty* it could never produce.

Acausality and Relativity

Einstein's *acausality* rears its head wherever collisions are missing in his analysis. Let us look as the 2nd ad hoc he used to install light as a particle: "Unlike other particles, it attained this velocity instantaneously when emitted from a source" (Table 3). Classical particles do not do this. A baseball, for instance, flattens slightly before it leaves the bat. On the contrary, disturbances in a medium are carried forth at relatively constant velocities commensurate with the interparticle velocities within that medium. Air, for example, transmits sound waves at 343 m/s, while the interparticle motion of the nitrogen molecules within is about 515 m/s.

Again, about the imagined light particle, the 3rd ad hoc states: "Unlike other particles, it would not take on the velocity of its source" (Table 3). If light really was a particle, this would be a violation of Newton's Second Law of Motion. If a baseball pitcher threw a baseball toward you at 100 mph in a convertible traveling 65 mph, you would be hit by a baseball traveling at 165 mph. De Sitter showed light was not a classical particle when he determined the velocity of light was the same for a dual star going away as for one coming toward him.[67] At this point regressives had to make a choice, either: 1) light is a wave in a medium or 2) it is a special particle that observes Einstein's ad hoc. The first was correct, but regressives chose the second.[68]

Einstein also violated *causality* when he implied that "Unlike other particles, light particles did not lose motion when they collided with other light particles" (Table 3). If light was a classical particle, it would indeed lose motion in collisions with other light particles. This is obvious, with all the light sources in the universe emitting light in all directions. Regressives would say these sources are too far apart for their emitted photons to collide with others. That is a particularly specious argument—by moving a millimeter left or right you still can see the same star equally well. That is because the aether medium transmits light waves in the same way the atmosphere transmits sound: as waves and not as particles.

---

[67] de Sitter, 1913, An astronomical proof for the constancy of the speed of light.
[68] Remember, all these regressive circumlocutions were required only because Michelson and Morley incorrectly assumed the ether medium they sought was not entrained around Earth.

Glenn Borchardt

CHAPTER 3: CERTAINTY

The Third Assumption of Religion

It is possible to know everything about some things.

Certainty is comforting. We crave predictable outcomes. *Certainty* is preached from every pulpit: There certainly is a god, a heaven, a hell, living after dying... Do *this* and your life will have ultimate meaning. Do *this* and you will live forever in a special place made for those who believe what I say.

The science evangelists, such as Neil deGrasse Tyson, spread the regressive paradigm as if it were certain, reminding us "Einstein is always right"—even though he is most often wrong.[69] We would love to have *certainty;* an anchor point we could grab hold of in a turbulent intellectual sea only partially filled with knowledge. Even when the theoretical necessity for aether became obvious, it was misconstrued by many as the imaginary ether, a fixed anchor around which all else must flow.[70] The search for *certainty* is endless in a universe that is endless.

Certainty is what Sheldrake was demanding in his attack on Newtonian mechanism: If you are as smart as you say you are, give us perfect precision, perfect predictions, and perfect knowledge about everything. That would never happen, of course, because the universe is

---

[69] As pointed out in our previous books, particularly Borchardt (2017).
[70] Again, this is what Michelson and Morley assumed in their famous 1887 experiment.

infinite. The opposing assumption is **uncertainty**, which states: "It is impossible to know everything about anything, but it is often possible to know more about anything." See how that only fits with *infinity*? Sheldrake's religiously based demand that there be an end to all the pluses or minuses, all the fuzziness, all the **uncertainty**, never can be realized because the universe is infinite. This leaves him with another outlet: causes that are not physical but are nonetheless able to effect change without material collisions. That actually makes *certainty* just another form of *immaterialism*. Like the regressives who do not know the physical cause of gravitation, without having to specify what is colliding with what, he can imagine anything he wishes. The "things" of his imagination do not even have to exist, unlike real things, which have XYZ dimensions and location with respect to other things and therefore consist of matter—his nemesis as an immaterialist.

Certainty and Religion

According to Dr. John Shook, "religious certainty is a dangerous weapon." In support of that unpopular claim, he writes:

> *People often enjoy the feeling of certainty. This pleasant emotional state is similar to other artificially induced euphorias. It can be addictive, and self-destructive. Worst of all, it can turn a person into a threat to the rest of society. ... And what if all this certainty affected their public behavior to the point where they try to tell others that they are wrong, and then they try to manipulate and control other people's lives? We don't have to speculate what would happen — this situation is called "Religion."*[71]

---

[71] Shook, 2010, Religious certainty is a dangerous weapon.

With over 4,200 dissimilar religions, that is a lot of world-wide *certainty*. No doubt most believers are *certain* they were lucky to be born to the "one true religion" and that the other 4,199 are false. The results have not been pretty: the world suffers from daily atrocities perpetrated by religions. As Voltaire famously wrote in 1765: "Those who can make you believe absurdities can make you commit atrocities." It seems the greater the contradictions between religions, the greater the reactions. Welcome to 2020.

Uncertainty and Science

Sorry, but there can be nothing certain in an Infinite Universe in which all things are in motion. Heisenberg realized this at least for the tiniest objects in devising his Uncertainty Principle. One might be able to determine either the position of a particle or its velocity, but not both at the same time. That is because, to do either determination, one must interact with that particle in some way, thereby changing both its position and its velocity. This has nothing to do with size because all microcosms work the same way. The experimenter is always unavoidably part of the experiment. For example, sociologists must minimize their intrusions in order to get useful data when studying a particular culture. All scientific disciplines require a viewpoint from which to select and interpret data that are theoretically infinite in number.

Certainty and Relativity

Certainty, finity, and *acausality*

Einstein's belief in certainty was manifest in his dispute with quantum mechanists. Like others imbued with the Newtonian tradition, he was a believer in finite universal causality. That means he thought there must be a finite number of causes for any particular effect—a proposition requiring his early, equally idealistic belief in perfectly empty space. The presence of a universal aether would interfere with the possibility of causality being finite. There always would be a plus or minus in any set of determinations for a particular effect no matter how small the microcosm.

That is what the quantum mechanists were running up against even though they did not seem to realize it. As in all the sciences, there are an infinite number of often insignificant causes unknown to us that affect the outcomes of our determinations per **uncertainty**. But in the face of aether denial, infinite universal causality (***causality***) was not a possibility at the time. And it still isn't: The Relativity-Quantum Mechanics Paradox is one of the big sore points for regressive physics. Here is how Einstein famously put it, tying together his early religious views with his religious assumption of certainty:

> *Quantum mechanics is certainly imposing. But an inner voice tells me that it is not yet the real thing. The theory says a lot, but does not really bring us any closer to the secret of the 'old one'. I, at any rate, am convinced that He is not playing at dice.*[72]

Let me speculate on what Einstein is voicing here. He is claiming, not only a belief in god, but just like Newton and Laplace, he is implying perfect prediction is possible. In objecting to the "playing dice part" he is de facto objecting to the ***infinity*** inherent in all measurements. A mere improvement in the model would never satisfy Einstein. As mentioned, effects would require a finite number of causes for us to produce perfect predictions, a completeness to theory that the Infinite Universe cannot provide. As a result, like snowflakes, no two predictions and no two results can be identical. All experiments must have a plus or minus "error." Not realizing this, religious folks like Sheldrake might smugly proclaim it as indicative of science's failure to discover complete and perfect truth. Per ***uncertainty***, this has nothing to do with throwing dice, and everything to do with our inability to know everything about anything. Even without

---

[72] Einstein, 1926, Letter to Max Born on quantum mechanics, December 4.

taking aether particles into account, any work performed in Earth's atmosphere must deal with experimental objects subject to trillions of collisions from nitrogen and oxygen molecules traveling at 515 m/s from all angles. It is impossible to consider all those causes and we are particularly fortunate and generally elated whenever the missing causes turn out to be insignificant.

Relativity-Quantum Mechanics Paradox

As mentioned, a paradox always has at least one incorrect assumption. In this case, it is the religious assumption of *finity* as alluded to above. As we will see, Einstein's belief in perfectly empty space required his unconsciously assuming all Ten Assumptions of Religion. Without empty space, his Untired Light Theory, based on his eight ad hocs, never would have resulted in Special or General Relativity Theory. The quantum mechanists assumed *finity* as well, but they resolved their problem with the Heisenberg Uncertainty Principle by inventing the Copenhagen Interpretation of quantum mechanics, which treats probability as a singular cause. By lumping the ***infinity*** of causal factors not discovered in any experiment, regressives kept their belief in the religious assumption of *finity* intact. By doing so they had no conflict with Einstein's empty space hypothesis. Without aether, however, any wave motion discovered had to be attributed to the objects themselves. That is how the "particles are made up of waves" trope got started.

The upshot is that in the battle between relativity and quantum mechanics, only quantum mechanics can survive. Aether denial and empty space is critical for relativity, but only an embarrassing nuisance for quantum mechanics. Both the Copenhagen Interpretation and wave-particle duality finally will be discarded when *finity* is replaced by ***infinity***. Quantum mechanics would be greatly improved with the application of univironmental determinism. As with all microcosms, the study of the infinite matter in motion in the environment is just as important as the infinite matter in motion within. That is not possible for relativity, with its massless-perfectly empty particle existing within a massless-perfectly empty environment. Einstein's attempt to turn wave motion into particle motion is revealed to be completely vacuous.

Glenn Borchardt
Certainty and the Constancy of the Velocity of Light

For more than a century we have been drilled with the claim the speed of light in vacuum is constant. This false, "revolutionary" assumption defied observations there are no constants in nature.[73] It really wasn't all that revolutionary, being founded on *finity*, which was a primary characteristic of classical mechanics. The reason there are no constants was alluded to above: The Infinite Universe cannot produce them. That makes Einstein's 1st ad hoc especially egregious: "Unlike other particles, his light particle always traveled at the same velocity—it never slowed down" (Table 3). He had to make that postulate to convert what were waves moving through aether into particles moving through empty space. The main problem: Perfectly empty space has never been found. Wave motion through a medium can be relatively constant over extreme distances, but solitary particles must eventually slow down as they collide with other things—especially other light particles. In other words, light particles would lose energy over distance. For particles, this would result in a decrease in velocity.

Wave transmittance through a medium also results in a loss of energy, but, because velocity through a medium is controlled by the properties of the medium, there is no decrease in velocity over distance. Voila! Thus, because light does not lose velocity over distance it cannot be a particle. Energy losses in a medium appear as redshifts, and that is exactly what is observed when light travels cosmic distances. It seems obvious to me that these "cosmological redshifts" occur because the reproduction of waves is imperfect. That is, a wave certainly cannot produce an exact, perfect, copy of itself. The imperfection appears as a lengthening of the distance between wave peaks. Hubble and others suspected as much, with their interpretation being labeled facetiously by regressives as the "Tired Light Theory." I devised the antidote to that one as Einstein's "Untired Light Theory,"

---

[73] Puetz and Borchardt, 2011, Universal Cycle Theory: Neomechanics of the Hierarchically Infinite Universe.

which pokes fun of the impossible empty space and perpetual motion required for his particle theory of light to work.[74]

I suppose an idealist could consider the velocity of wave travel through the aetherial medium to be constant. But here we cannot be so naïve. Because all things in the universe are continually in motion, the properties of all things, including media, continually change. For instance, the density of the aether medium increases with nearness to Earth while its pressure decreases.[75] Even in outer space, we do not expect aether to be any more homogeneous than the rest of the universe. The velocity of light, like the velocity of sound, is relatively constant but, like sound velocity, light velocity never could be perfectly constant.

Certainty and the Doppler Effect

The fact distant galaxies exhibit redshifts as a function of distance has been interpreted as certain evidence for galactic recession. This is touted by cosmogonists as positive proof the universe is expanding. Unfortunately, the required Doppler effects occur only in media. Light traveling as a particle through Einstein's perfectly empty space could not exhibit a redshift. Realizing this, regressives made up the "wave-packet" model of the photon in a dubious attempt to handle the contradiction. Otherwise known also as "wave-particle duality," we are supposed to believe the photon brings its own waves along with it as it travels through the universe. This kind of thinking has gotten so out of hand that quantum mechanists and some reformists have imagined all things in the universe to consist of waves. Now, waves are motions, not things, as will be discussed in the chapter on *separability*. The idea that a single particle consisting of nothing (i.e., having no mass), traveling through empty space could carry waves of matterless motion along with it is not only absurd, it is vacuous.

---

[74] Borchardt, 2017, Infinite Universe Theory.
[75] Borchardt, 2018, The physical cause of gravitation.

Glenn Borchardt

CHAPTER 4: SEPARABILITY

The Fourth Assumption of Religion

Motion can occur without matter and matter can exist without motion.

Separability, like *immaterialism*, forms another of the least subtle assumptions of religion. Both are direct contradictions of the mechanist's claim there are only two basic phenomena characteristic of the universe: matter and its motion. Theologians, however, would not be theologians unless they lobbied for the idea "there is more than just matter and motion in the universe." Exactly what that could be without it being matter and the motion of matter is impossible to describe. In all languages, descriptions involve nouns (matter) and verbs (motion). Even those who imagine and speak of ghosts or spirits must imagine some XYZ portion of the universe that is in a particular location with respect to other portions. Those ghosts or spirits are meaningless to the imagination unless they move. They have to *do* something to get our attention. The imagined ghost in the attic at least has to make a noise when the house cools. Otherwise the "haunted" designation would have be removed from the tourist brochure.

Separability and Religion

Immaterial, imagined gods traditionally consisted of motion without matter. That is because imagined "things" do not exist—they contain no matter and do not have XYZ dimensions. What is left is the tendency to objectify motion, as Einstein did with light. The matter (wind) that moves tree branches cannot be seen; the matter (electrons) in the lightning bolt cannot be seen. The universe has an infinite number of motions so, early on, there were a lot of gods. As the causes of some of these motions became better understood, the number of gods declined. Most professional theologians probably consider getting down to three or one as a sign of religious progress. After all, the fewer superstitions you have, the better you can deal with reality—the nemesis of every religion. Still, religious folks might have problems ascribing materiality to any of their gods. Being an XYZ portion of the universe and having location with respect to other portions just does not seem to be in the cards.

It is ironic that Georg Wilhelm Friedrich Hegel (1770-1831), the German idealist philosopher who proposed *inseparability*, was a famous dualist. That is, he considered the universe to consist of matter and spirit. This was in spite of the famous quote establishing our Fourth Assumption of Science:

> *Just as there is no Motion without Matter, so too, there is no Matter without Motion.*[76]

This wise dictum is taken somewhat out of context. Except for his introduction of dialectics, much of Hegel's stuff, like that of other indeterministic philosophers, was a jumbled, illogical mess. You only have to read the page that quote was taken from to prove that to yourself. His work apparently was a take-off on Newton's laws of motion, which were highly successful in establishing mechanics and its philosophical counterpart, mechanism, which blatantly assumed the universe only consisted of matter in motion.

The Hegelian saga has a lesson for us. Despite his partial adoption of scientific mechanism, Hegel's inclusion of the spirit idea made it widely popular among those following the religious traditions of the time. Though only 30 years old when he wrote at the beginning of the 19th century, like Einstein, he produced a mixture of science and religion suited to the times, getting famous in the process. Of course, the mixture at the beginning of

---

[76] Houlgate, 1998, The Hegel Reader, p. 270.

the 20th century had to be much more subtle if it was to be accepted as "scientific." There was no overt injection of spirits and other overt claims of religion into relativity, despite its use in support of various religious notions. That is why we need the present book. We need to carefully examine the religious suppositions Einstein unconsciously used in developing relativity. Otherwise, we would be unable to understand why leading physicists and cosmologists came to the amazing conclusion that the entire universe exploded out of nothing.

Disincarnation

Per Gould's NOMA, religion is a magisteria that is not supposed to cross the line into the scientific magisteria. Ideas originating in ancient dreams and imaginings are supposed to stay there. Unfortunately, that is almost never the case. Theologians and their followers often speak of "souls" and "spirits" as though they really existed. They cross the line because the external world is replete with examples of the motion of matter. Again, matter exists; motion occurs. Matter has XYZ dimensions; motion does not. Not being fully cognizant of these facts, they incorrectly assume *separability*. With that foundational religious assumption, they then come up with the common mistaken idea the motion of the body can be separated from the body (i.e., disincarnation). Thus, an active live animal is said to be "spirited," while a dead one is not.

Where did that spirit go when that animal died? The answer to this question is the scene of what I call Einstein's greatest blunder[77] and the only reason relativity was possible.[78] When we ask that question in the way theologians do, we are doing what Einstein did: objectifying motion. We are thinking of motion considered as spirit or soul as a thing, an object that can be separated from the body. Our dreams reinforce this whenever our

---

[77] This blunder is not what you were taught. What Einstein admitted to was trivial in comparison.
[78] Borchardt, 2011, Einstein's most important philosophical error.

dead friends and relatives seem to appear very active in them. Although some religions do not believe in life after death, the ones that do invariably assume the 4th assumption of religion: *separability*.

Inseparability and Science

Remember again the opposing assumption of ***inseparability***: "Just as there is no motion without matter, so there is no matter without motion." This is the foundational supposition of mechanics and Newton's three laws of motion. It is what made Isaac Newton (1642-1727) the greatest scientist who ever lived. You might wonder why Newton did not bring his main supposition into the light of day by proclaiming it a fundamental assumption. That was because his supposition was a "presupposition," an unstated assumption that is part of our unconscious. Presuppositions are never stated. We all have them all the time. For instance, we usually can walk across the floor without first stating formally that we assume the floor will support our weight.

Despite the outstanding success of ***inseparability*** in mechanics, its obvious contradiction with *separability* has made it non grata to religious folks. Nowadays it is generally avoided in scientific texts, with the term "energy" being used instead whenever possible. That is one reason regressives state the First Law of Thermodynamics (***conservation***) in terms of energy instead of matter and motion. The problem for scientific philosophy is that energy does not exist—it is a calculation. On the other hand, matter clearly exists and motion clearly occurs. Implying that energy neither can be created nor destroyed is the same as saying a calculation neither can be created nor destroyed. Physics is replete with such calculations of matter-motion terms, which are indispensable but nonetheless tend to be improperly objectified during regressive interpretations.

Separability and Relativity

Had Einstein used Hegel's dictum stating ***inseparability***, relativity never would have plagued the globe. Again, his use of the opposing, incorrect assumption, is quite subtle even though it results in the shaking of wiser heads to this day.

Objectification of Motion

Einstein's objectification of motion is plainly seen in his 6th ad hoc: "Unlike other particles, any measurement indicating light speed was not constant had to be attributed to 'time dilation'." And more broadly in his 7th ad hoc "Time had to be considered something other than motion, for motion cannot dilate." Like so many others, Einstein did not say what time is. To this day, and partly due to Einstein, most folks do not know what time is either. Some think it is a mystery, a measurement, an illusion, or a calculation.[79] Lee Coe explained it simply: "time is a property of matter, and does not exist except as a property of matter."[80] This is close, but no cigar. Time is motion. Strictly speaking, it does not "exist," it occurs. This is an extremely important quibble, because we define existence as an XYZ portion of the universe having location with respect to other portions. My esteemed colleague, Steve Puetz, once thought "time was an aspect of motion." As I explained in one of our most popular Blogs,[81] which was stimulated by his comment, motion is an abstraction for what all things do. There are an infinite number of types of motion just as there are an infinite number of types of matter. However, unlike material things, those motions do not have aspects.

According to Oxford, an aspect is "a particular part or feature of something." The "part" and "something" refer to XYZ portions of the universe. Time (motion), however, is not an XYZ portion of the universe; only the thing moving is an XYZ portion of the universe. True, motion must occur *within* a portion of the universe, but that motion is not a "something." Anyone, like Einstein, who considers it such commits the sin of objectification, even if the consideration is only of one dimension, as he does in General Relativity Theory. Dimensionality does not apply to time.

---

[79] Gisin, 2020, Mathematical languages shape our understanding of time in physics.
[80] Coe, 1969, The nature of time.
[81] Time is Motion

Anyone who treats time as an object commits the sin of violating *inseparability*.

Most importantly, motion (time) cannot be combined with matter. We cannot attach a piece of motion to a piece of matter. Matter and the motion of matter are not separable in this way. Motion is what matter *does*. Like the elder Einstein, I have argued that space is material. But if time is motion, time cannot be combined with space. Time (motion) is what space does. Unlike Einstein, I therefore consider his "space-time," like much of his work, to be purely imaginary. Perhaps this is why Einstein once admitted:

> *I am enough of an artist to draw freely upon my imagination. Imagination is more important than knowledge. Knowledge is limited. Imagination encircles the world.*[82]

Indeed, that "imagination encircling the world" seems a veiled apropos bow to globally ubiquitous religions, which are invariably and gloriously founded on dreams and imagination. In science, however, imagination and knowledge are supposed to have a reciprocal relationship. The scientific imagination is constrained by a special requirement: to interact with the external world through observation and experiment. Space-time can have no physical existence; we cannot mix a little matter with a little motion and come up with anything real. In General Relativity Theory we can do some math and envision extra-Euclidean dimensions, but it is all just imaginary which, like religion itself, then becomes more important to the practitioner than reality.

The successes attributed to imaginary space-time in General Relativity Theory result from the material constituents of space, whether they result from refraction due to an atmosphere or a medium conducting light waves and shock waves.[83]

---

[82] Viereck, 1929, What life means to Einstein.

In Special Relativity Theory we can do some math and envision time as a malleable object that can dilate just enough to satisfy our equations, but it is all just imaginary which, for Einstein, then becomes truly more important than reality. None of this is possible without the religious assumption of *separability*. Time, like the imagined human spirit, is thought of as an object that could be separated from its material carrier in a sort of disincarnation. Einstein managed this separation in his famous 1905 paper. All he had to do was substitute one little letter, t, for another, l.[84] Although we often use time to measure distance, time is not distance. Thus, if we travel the 100 km from point A to point B in one hour, the distance is still 100 km. The distance is not one hour. This subtle error apparently was not caught during peer review and has been accepted by regressive physicists ever since. This little sleight of hand in a bow to *separability* was only one of Einstein's disincarnated ghosts.

Separability and the $E=mc^2$ Equation

Another was his treatment of the $E=mc^2$ equation, which actually was devised by Maxwell in 1862.[85] This perfectly good equation does not depend on relativity, although Einstein's promotion of it was what got him famous, with each successful use contributing to his notoriety. It was not until 1950 that he admitted he got it from Maxwell *without attribution*.[86] Einstein and his followers generally present the equation as proof that mass and energy are equivalent. To this day, even NASA thinks the universe consists of 4.84% baryonic matter, 25.8% dark matter, and 69.2% dark energy. While dark matter is probably decelerated aether, energy cannot possibly exist. That is because "energy" is a calculation, and calculations cannot do anything. The equations for energy describe the motion of matter. The energy NASA refers to was invented to save the Big Bang Theory. With improved telescopes looking at increasingly distant galaxies,

---

[83] Borchardt, 2017, Infinite Universe Theory.
[84] Bryant and Borchardt, 2011, Failure of the relativistic hypercone derivation.
[85] Ricker, 2015, The origin of the equation $E=mc^2$.
[86] This is a cardinal sin among scientists. Today's reference lists can double the length of a paper.

cosmological redshifts increased so much that the Hubble equation indicated those distant galaxies were receding at greater than light speeds. Per relativity, that would never do. Extra energy was needed to fuel the calculated inflation of the empty space between supposedly receding galaxies. This imaginary energy has no matter associated with it—above all not the "dark matter" that is well-established and tends to form an "aetherosphere" around cosmological bodies. Note, I have proposed that dark matter is decelerated aether that accumulates around baryonic matter after it does the little acceleration job that causes gravitation.[87] The $E=mc^2$ equation itself tells the story. If there was no matter (m) associated with dark energy, $E=mc^2$ would be 0. The dark energy ad hoc without matter amounts to matterless motion—a clear violation of *inseparability* and an affront to anyone with any sense. That has not deterred folks with a religious bent. The imaginary "energy" leaving the scene of the $E=mc^2$ event flits away like a ghost, a massless chimera best suited to a pseudo-intellectual Sunday sermon.

Again, there can be no motion described by energy without its material carrier. Thus, the energy of the falling water (matter in motion) turns the turbine (matter in motion) that turns the magnets that generate electricity (matter in motion). However, the usual regressive interpretation of the $E=mc^2$ equation claims mass and energy are equivalent. In particular, regressives say mass can be turned into energy and energy can be turned into mass. That is not the way it works. There is no such thing as "energy" per se. You cannot give me a bunch of energy. All you could do would be to present me with access to a flowing waterfall, a turbine to transfer the motion of falling water to the rotating motion of the turbine, and magnets to get those electrons moving enough to be useful as electricity.

The Physical Meaning of $E=mc^2$

As I explained over a decade ago, the correct interpretation of the $E=mc^2$ equation requires strict adherence to *inseparability* and the theoretical necessity for aether.[88] You can get the details from that paper

---

[87] Borchardt, 2018, The physical cause of gravitation.

and from "Infinite Universe Theory." Here is a summary. Take a look at Figure 5, which is a simple demonstration of Newton's Second Law of Motion with regard to the $E=mc^2$ equation. Here I show a portion of the universe, which I designate a microcosm. Within that portion are smaller portions I designate submicrocosms and outside the microcosm are other microcosms I designate as supermicrocosms. The figure shows the usual second law collisions described by the F=ma equation. In the first half of the figure, collisions by supermicrocosms accelerate submicrocosms; in the second half, collisions by submicrocosms accelerate supermicrocosms. The necessarily irregular boundary of the microcosm allows this transfer of motion. What we are seeing here is the absorption of motion in the first case and emission of motion in the second. If we removed the microcosm entirely all we would have left would be described by Newton's First Law of Motion: a supermicrocosm traveling from left to right under its own inertial motion. Per Newton, the material body on the left would continue to be the same material body when it was on the right. It would not change from matter in motion into a calculation or mysterious entity called "energy."

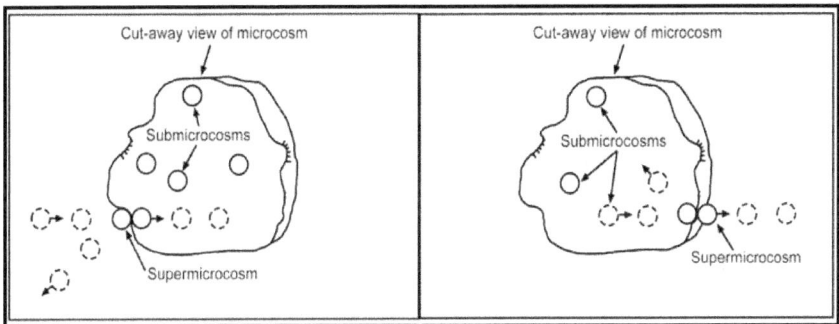

Figure 5. **Neomechanical** interactions apropos the $E=mc^2$ equation illustrating that both absorption and emission involve mechanical collisions described by Newton's Second Law of Motion. By denying aether exists, regressive physics denies that these collisions occur. Credit: Borchardt (2017, figure 25).

---

[88] Borchardt, 2009, The physical meaning of $E=mc^2$.

Now, Maxwell was particularly concerned with the absorption part of the reaction when he discovered the equation in 1862, presenting it as m= E/c^2, which is a simple rearrangement of $E=mc^2$. He was interested in the reason things (microcosms) gained mass when exposed to sunlight or heat. The emission part of the reaction was noted by science evangelist Martin Gardner who wrote: "As the coffee cools, mass is lost."[89] That was correct even though Martin tried to explain it by using relativity and its emission via Einstein's magical massless light particles—a typical einsteinism.

But, as implied above, the changes in mass described by the $E=mc^2$ equation can be explained simply by Newtonian mechanics—relativity and its religious assumption of *separability* are unneeded. What *is* needed are the scientific assumptions of **inseparability** and **infinity**. These assume motion needs matter for its transference, that matter is infinitely subdividable, and that aether theoretically must exist. Remember mass is defined as the resistance to acceleration. Supermicrocosms accelerate submicrocosms via collisions (i.e., $F=ma$), producing increases in the momenta (i.e., $P=mv$) of the submicrocosms. As in my football team example,[90] the more active the constituents of a microcosm, the harder it is to counteract their accelerated impacts. This appears as an increase in mass, with the reverse occurring when those submicrocosms accelerate other supermicrocosms through the microcosmic boundary. In other words, when internal motion increases, mass increases; when internal motion decreases, mass decreases.

Again, the key to this explanation remains the necessity for **inseparability**: "Just as there is no motion without matter, so there is no matter without motion." For any of these changes in mass to occur, there must be supermicrocosms in the macrocosm. Massless particles will not do; perfectly solid particles will not do; a calculation (energy) will not do. Regressives who think relativity is required for explaining changes in mass may not realize it, but they are using two religious assumptions to do so: *separability* and *finity*.

---

[89] Gardner, 1962, Relativity for the Million, p. 66.
[90] Borchardt, 2017, Infinite Universe Theory.

Glenn Borchardt

The Twin Paradox

This contradiction, founded on *separability*, probably amazes more beginning physics students than any other. Those gullible enough to swallow it hook-line-and-sinker can expect to pass the course. They might even go on to become the next cosmogonist to prepare the next ad hoc that momentarily extends the life of the tattered Big Bang Theory. None of this would still be occurring today if Einstein had known just a little bit of scientific philosophy: A paradox always means at least one of your assumptions is wrong.

By now, the twin paradox has become part of popular mythology. According to Einstein's equations, the inperiment goes like this: One identical twin leaves Earth in a high-speed rocket ship. Upon his return, he thinks his twin brother is much younger than he is. The stay-at-home twin thinks just the opposite, that the traveling twin is younger.

Remember what I said before about the nature of time. Time is motion. Universal time is the motion of everything with respect to everything else. We could never measure that. Instead, we only can measure specific time by using a device that records cyclic motion: a clock.

According to Steven Bryant,[91] the resolution of the Twin Paradox simply involves averaging the two Doppler shifts observed by the traveling twin and the Earth-bound twin. In his train example, Bryant nicely describes what actually is happening. He imagines two equally solipsistic observers, each with frequency counters. That is, each measures time from his own perspective. In Case 1, both agree on the frequency when the train is not moving. But:

> Now consider Case 2, where the train is moving away from the station at velocity v. Person A, at the train station, blows his horn, which is picked up on both

---

[91] Bryant, 2011, The twin paradox: Why it is required by relativity.

frequency counters. However, due to the Doppler shift and resulting lower frequency observed on the train, the frequency counter on the train is incrementing more slowly than the frequency counter at the station. In other words, the frequency counter on the train will take longer to reach 1000 beats than its counterpart at the station: The clock on the Train is running slower.

As a continuation of Case 2, consider the situation where Person B, who is on the train, blows his horn. The sound is picked up on both frequency counters. In this case, again due to the Doppler shift, the frequency counter at the station will run slower than its counterpart on the train. Once again, a wavelength-based model explains this using Doppler shifts, while Einstein's length-based theory explains it as time dilation.

During the twin's return trip, the situation is reversed, with the Doppler shifts being shorter for both twins. The upshot is that both twins have the same age once they agree on which clock to use. The correct clock, of course, is the one that measures time when they are both together. The resolution of the paradox has nothing to do with relativity. Any time Einstein and his followers call for "time dilation," you can be sure their measurements fail to use the Lorentz Correction Factor properly. Again, time is motion. Motion cannot dilate, only things having XYZ dimensions can dilate. Again, time dilation is based on the religious assumption of *separability.*

The reason relativists hold so fast to such nonsense involves Einstein's insistence that the velocity of light is magically constant for all observers. But like all wave motion within a medium, wave velocity is constant only with respect to that medium. Obviously, results vary when velocity measurements are taken from different viewpoints (i.e., "reference frames"). Thus, if you wished to measure wave velocity in a body of water, you would have to take your own motion into account when calculating the velocity. When light is incorrectly considered a particle instead of a wave in the aether, varying amounts of "time dilation" would be needed to calculate the correct velocity for c. Einstein's claim aether is irrelevant in relativity is correct—as long as you can stomach time dilation and the rest of his egregious, unprecedented ad hocs. You have a choice: aether or time dilation. Reformists be careful: dumping time dilation dumps relativity along with the twin paradox.

Religious Roots of Relativity

Glenn Borchardt

CHAPTER 5: CREATION

The Fifth Assumption of Religion

Matter and motion can be created out of nothing.

While most religious ideas are descended from dreams, the religious assumption of *creation* isn't necessarily one of them. Our activities in the external world require us to seek the physical causes for effects. We create things by bringing other things together. We may imagine what our prospective creations may be like, but even the most besotted engineer or artist cannot create something out of nothing. Based on our experiences, it is logical to assume that if everything we know has a cause, that the universe must have one too.

Creation and Religion

Creation stories, of course, are critical components of most religions. You don't have to be religious to wonder why there is something rather than nothing. "Where did it all come from?" is one of the BIG questions the religious Templeton Foundation tries to answer by perennially dealing cash to scientists who will accept the bribe. Religious folks like to imagine there once was nothing and that the nothingness suddenly was filled with everything we can see around us. Most religions have dreamed up a creator they think must have done this. Thousands of years ago that job was thought to have been performed in six days according to the Bible. But that was when Earth was flat, the Sun went around it, and the stars were attached to a celestial sphere also created just for us. Now, we have our own little Milky Way galaxy, but there are over two trillion other galaxies like it that we can see, making the creation job quite an effort for a single creator. The grandeur of Einstein's "old one" truly is all-encompassing, at least for anyone who has examined the real world in some detail since Einstein.

Creation versus Evolution

But religions traditionally have discouraged such examination ([Figure 6](#)). The Pope is right. Anytime we take our religious dreams and imaginings into the external world we invariably will undergo confusion and cognitive dissonance. This is especially obvious in the *creation* versus evolution debate still raging in the US. Although, the Pope has been loosening up on evolution,[92] evangelicals are particularly outraged by the usual claims by biologists that the theory of evolution is critical in their work. The latest version of creation is called "intelligent design," with some of the slightly more educated promulgators even suggesting the all-powerful designer did not create everything all at once, but acts through evolution to do so. Promoters can imagine a magical intelligent designer, but assuming *finity*, cannot imagine how so much variety and complexity could have formed at the time of creation.

It might be okay if creationists stuck to the religious part of Gould's [NOMA](#), but the macrocosm keeps intruding, as it always does. When we obtain ever-increasing objective knowledge about our surroundings, religious "knowledge" derived from dreams takes a beating. Those with eyes open must account for every fossil and its position in the landscape and for every decay of every radioactive isotope. While the Catholics may have given up, the evangelists still try to cram all of universal history into the 6,000 years implied by their "infallible" text. About the most ridiculous is the 500' exact replica of Noah's ark in Kentucky in which dinosaurs are depicted alongside humans ([Figure 7](#)).

---

[92] McKenna, 2014, Pope says evolution, Big Bang are real.

And the spirit of curiosity is not a good spirit. It is the spirit of dispersion, of distancing oneself from God, the spirit of talking too much. And Jesus also tells us something interesting: this spirit of curiosity, which is worldly, leads us to confusion.

—Pope Francis

Figure 6. Pope discourages curiosity. Credit: Jerry Coyne.

Figure 7. Dinosaurs in Noah's ark at the **Ark Encounter**, a Christian theme park funded, in part, by the State of Kentucky. These are supposed to be models of Baryonyx, which became extinct about 125 million years before humans evolved. Credit: **Peggy Watson, Citizen of Heaven, Follower of Jesus**.

Conservation and Science

With regard to the scientific assumption of **conservation**, I am reminded of the time I was assigned by the State of California to discover evidence for earthquakes that occurred before 1849. Before the gold rush, there were no newspapers to report damage we could use to determine size and location of events before we had seismometers. Pioneer letters, diaries, manuscripts, and mission records often had valuable damage reports. During my reading I came upon this amusing recollection of a priest in 19th century California who was trying to convert one of the natives. When the subject of the origin of the universe came up, the priest gave his usual Genesis story: "You know that God created the universe, don't you?" The native replied: "Why you silly man! The earth has always been here!" Obviously, belief in *creation* was not universal among the natives—at least until the church took over. The wise native was declaring a certain permanence to things, which is expressed formally in our assumption of *conservation*.

Conservation is the Fifth Assumption of Science: "Matter and the motion of matter can be neither created nor destroyed." It is derived from the First Law of Thermodynamics, normally considered by regressives as the Law of Conservation of Energy, which states "Energy can be neither created nor destroyed." Because energy does not exist (remember, it is a calculation) I had to state the assumption in mechanical terms. In either rendition, such statements are direct contradictions of the religious assumption of *creation*.

Regressives and reformists generally think of evolution as the opposite of *creation*, traditionally focusing on biology as I did in the previous section. We have bigger fish to fry, however. In "The Scientific Worldview" I proposed univironmental determinism, as the universal mechanism of evolution (the observation that what happens to a portion of the universe is determined by the infinite matter in motion within and without). One of the subtle appeals held by the Big Bang Theory is its claim to be evolutionary, despite its seldom-acknowledged foundation on the assumption of *creation*. Perhaps that is why really smart evolutionary biologists, such as Professor Emeritus Jerry Coyne of the University of

Chicago, have no objection to absurd claims the universe exploded out of nothing. In handling that contradiction, maybe Gould would have said: "Give us our evolution magisteria in biology and we won't bother your *creation* magisteria in cosmology."

Creation and Relativity

Although there are numerous miraculous claims within relativity, Einstein, like most scientists, did not call upon *creation* directly in his work. For example, substituting l for t may be creative sleight of hand,[93] but it is not easily recognized by regressives as being founded on *creation*. Of course, that is what it amounts to. The substitution of length for motion is the substitution of something for nothing. Jumping then to General Relativity Theory was easy. Time could be included with the three dimensions as if it was material. The result was 4-dimensional space-time, which, like religion itself, was purely imaginary. Folks like myself never were privileged to observe a god or four dimensions no matter how much we tried. Mathematicians such as Riemann and Friedmann had no problem with that.[94] Math is great for describing reality, but it is not always a reliable test of it.

Empty Space and Space-time Rescue Creation

Einstein spread bogus space-time all over the universe in tune with his aforementioned delusion that imagination was more important than knowledge. Also as mentioned, he did the same with his invention of his massless light particle and the perfectly empty space needed for its unprecedented perpetual motion through it. These religious assumptions

---

[93] Bryant and Borchardt, 2011, Failure of the relativistic hypercone derivation.
[94] See, 1925, Researches in Non-Euclidian Geometry and the Theory of Relativity: A Systematic Study of Twenty Fallacies in the Geometry of Riemann, Including the So-called Curvature of Space and Radius of World Curvature, and of Eighty Errors in the Physical Theories of Einstein and Eddington, Showing the Complete Collapse of the Theory of Relativity.

provided the critical foundation for *creation*'s grand finale: The Big Bang Theory.

The explosion of the entire universe out of nothing boggles the mind. Previous *creation* myths were trifling in comparison to the grandeur of its scope and the arrogance and *certainty* with which it is presented to the public. The *creation* of Earth and its accoutrements out of nothing at least had an imaginary god responsible for the process. But two trillion galaxies with nothing but imaginary "dark energy" as the culprit? I have given a few hints about how we got to this point. Let me reiterate the recipe complete with the religious assumptions required:

1. Imagine a massless light particle (*immaterialism, finity*).
2. Imagine perfectly empty space (*immaterialism, finity, absolutism*).
3. Imagine motion and matter can be combined as a wave-particle (*separability*).
4. Imagine the Doppler Effect can occur without a medium (*immaterialism*).
5. Imagine the cosmological redshift has only one cause: galactic recession or, when needed, the expansion of empty space (*acausality, absolutism*).
6. Imagine the universe is expanding in all directions at once (*finity, noncomplementarity*).
7. Imagine matter can be combined with motion as 4-dimensional space-time (*separability*).
8. Imagine things can form by coming apart (*noncomplementarity, acausality*).
9. Imagine the universe had a beginning (*finity, acausality, immaterialism, creation*).

As you can see, the Big Bang Theory requires a lot of imagining based on religious assumptions shown in the parentheses above. Most are a step removed from the kind of dreams religious folks might have. For instance, I doubt many people actually dream about perfectly empty space exploding into a finite, continually expanding universe. After all, to dream of "nothing" is to not dream at all. In "Infinite Universe Theory" I listed the objections to Big Bang Theory (Table 4).

Table 4. Falsifications, contradictions, and paradoxes disproving the Big Bang Theory—a review (modified from Borchardt, 2017).

Big Bang Theory Predictions — *Actual Observations*

Big Bang Theory Predictions	Actual Observations
The Big Bang Theory predicts that we should observe only young cosmological objects at great distances.	Instead, we see elderly galaxies and galaxy clusters at the limit of observation.[95]
In actual explosions, objects are scattered in all directions and do not collide.	Cosmological objects often collide.
The opinion that the universe is expanding is dependent on the "Untired Light Theory," which assumes that light can travel great distances without losing energy.	Nothing we know of can travel from one place to another without losing energy.
The universe supposedly exploded out of nothing.	That is a contradiction of the First Law of Thermodynamics, otherwise known as the conservation of energy.
The Doppler Effect, is/was considered responsible for the Cosmological redshift and the interpretation that most galaxies are receding from us.	The Doppler Effect only occurs in a medium. Einstein's corpuscular theory of light, denies the presence of a medium.
The space-time concept, as used in General Relativity Theory and Big Bang Theory, assumes time to be a dimension.	Einstein's objectification of time is invalid. Time is not an object; time is motion. The universe is 3-dimensional, just like everything we observe. "Time dilation" and other Einsteinian fantasies are products of aether denial and failure to use the Lorentz Correction Factor properly.
The Big Bang Theory is based on the assumption of *finity*.	The most plausible assumption is ***infinity***. There are over two trillion galaxies in the observable universe with no end in sight. An Infinite Universe cannot expand, because it is already full.
Big Bang Theory assumes ideal perfectly empty space is possible.	The existence of the universe implies that nonexistence (empty space) is impossible. There is no evidence for perfectly empty space.

---

[95] Gary, 2011, Astronomers find old heads in a young crowd; Vergano, 2014, Hubble Reveals Universe's Oldest Galaxies; Wall, 2020, Rare monster galaxy.

CHAPTER 6: NONCOMPLEMENTARITY

The Sixth Assumption of Religion

All things are subject to divergence from all other things.

Big Bang Theory reeks with *noncomplementarity*. Note, that in this religious assumption, divergence does not involve "some" things, but "all other things." That is what an explosion amounts to (Figure 8). Now let us explore what makes *noncomplementarity* a religious assumption.

Religious Roots of Relativity

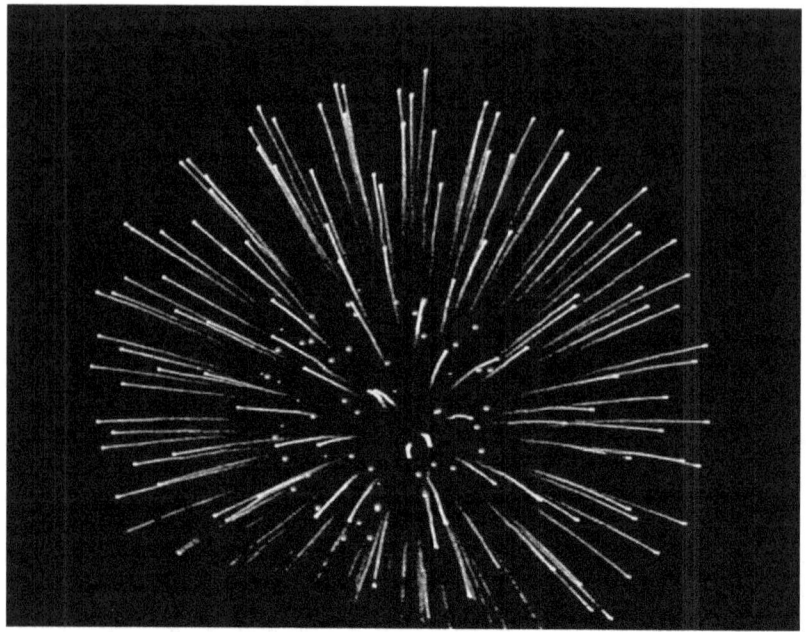

Figure 8. An explosion has only divergence, while creation requires convergence. Credit: Pinterest.

Noncomplementarity and Religion

This connection of *noncomplementarity* with religion is perhaps the most subtle of all Ten Assumptions of Religion. Be reminded that **complementarity** assumes there is a complement to the Second Law of Thermodynamics (SLT).[96] Books touting *creation* by some all-powerful god often include predictions in which that same god eventually will destroy his *creation*. Noah's flood came close. Revelation speaks of the rapture, in which all nasty material things will be destroyed. The Cold-War generation thought sure a nuclear war was going to destroy us all. Today's youngsters opposed to predicted horrible changes seek to "save the planet" before it is destroyed. Old timers contemplating their imminent physical

---

[96] Borchardt, 2008, Resolution of the SLT-order paradox. [The SLT states that isolated systems can only come apart; while its complement states that things come together to the degree that they are not isolated, such as in the Infinite Universe.]

divergence sometimes voice an opinion like this: "The world is going to hell in a handbasket!" *Noncomplementarity*, like the imagined hell, is the pessimist's dream.

Complementarity and Science

In opposition to *noncomplementarity* we have the Sixth assumption of science, **complementarity** which proclaims "All things are subject to divergence and convergence from other things." It is the missing dialectical complement to the Second Law of Thermodynamics. At this point I should give a bit of history about how I discovered this assumption.

History of Complementarity

I suppose you could say it stems from April 22, 1970 when the environmental movement got underway on "Earth Day." For too long self-centered humanity had been thinking more of itself than of the environment in which it lived. With burgeoning populations, much of the Earth was becoming a poisonous mess. In doing science, we tend to isolate portions of the universe in an effort to obtain useful predictions. We called these portions "systems." As mentioned, the regressive theory supporting that method is called systems philosophy, with which I eventually became infuriated. You see, I was trained as a soil scientist. Soils are so clearly products of rocks and the environments in which they exist. Systems philosophy tends to emphasize the system while slighting the environment. Soils were not like that. For instance, without precipitation from the environment, no soil. Without a reduction in the heat and pressure that formed the rock, no soil. This becomes especially obvious when you get out of the lab or office to get your hands dirty in the field. Eventually, I came to the conclusion both the system and its environment were equally important.

That's how I came up with univironmental determinism (UD), the observation that what happens to a portion of the universe is determined by the infinite matter in motion within and without.[97] Turns out this was the universal mechanism of evolution. It applies to all things, not just to biology. It is the proper replacement for Neo-Darwinism, which is only a special case of UD. Finally, we had the mechanism by which soils evolved, along with planets, galaxies, and humans as well.

Systems philosophy produced some particularly heinous regressive interpretations. One involved the Second Law of Thermodynamics. It got so bad that some out-of-the box thinkers began writing about the "heat death of the universe." In "Infinite Universe Theory" I wrote:

> *The Second Law of Thermodynamics states that all isolated systems eventually run down. In the regressive interpretation, constituent matter supposedly is converted into "energy," which escapes the isolated system as unusable heat. Another way of stating it from a mechanical viewpoint is that the constituents of the system will diverge or expand into its surroundings via their own momenta. Either way, both interpretations fit the expanding universe of the Big Bang Theory with divergence being assumed greater than convergence.*

On the contrary, in Infinite Universe Theory we assume all things are either going away from or coming toward other things. That is simply nature's dialectics. If we assume divergence and convergence are equal, we get **complementarity**, which is an automatic negation of the Big Bang Theory and the religious assumption of *finity* and upon which it is founded. And as I showed in one of my most notable papers, **complementarity** works on a less than universal scale as well.[98] Even before systems theory, mainstream science has always taught the Second Law of Thermodynamics as a divergence without its complement: convergence. A paradox then arises: If entropy and disorder always is increasing, how come we see so much around us in which entropy and disorder is decreasing? In

---

[97] Borchardt, 2007, The Scientific Worldview.
[98] Borchardt, 2008, Resolution of the SLT-order paradox.

mechanical terms, for everything that is coming apart, how come yet another thing is coming together? The correct answer is that the universe is infinite. Per *conservation*, the parts and the internal motions of a disintegrating object do not simply disappear. They continue through the universe, converging upon other things to form new objects. The SLT-order paradox, like all paradoxes, is founded on an incorrect assumption: *noncomplementarity*, the view that the "isolated" system is more important than its environment.

Noncomplementarity and Relativity

Einstein was agnostic on the question of whether the universe was infinite or finite. But as mentioned above, his "Untired Light Theory" led directly to the ridiculous interpretation the universe was expanding. An infinite universe has nowhere to expand into.

This is where the "heat death of the universe" trope came into play. Assuming the finite universe existed within a completely empty space, it also could be assumed to be an isolated system. Like all other systems so assumed, the finite universe would eventually undergo increasing entropy and disorder. It would run down like a clock isolated from the caretaker who replaces the batteries when needed. The "heat death" part was based on the usual regressive misinterpretation of the $E=mc^2$ equation. The correct interpretation requires aether (Figure 5). Completely empty space will not do. Heat is not some magical, matterless motion conjured by believers in the religious assumption of *separability*. Heat is vibration, and in this case, its escape from a finite, isolated universe nevertheless would require it to leave via vibrations in the aether medium. Like all real systems, perfect isolation cannot exist. There always is an environment of equal importance per *complementarity*, *infinity*, and univironmental determinism.

Religious Roots of Relativity

Glenn Borchardt

CHAPTER 7: REVERSIBILITY

The Seventh Assumption of Religion

Some processes are reversible.

Reversibility is common in science fiction, useful for dreaming about past events, and for impressing students by reversing chemical reactions in the lab. But is it really a religious assumption?

Reversibility and Religion

Even in religion *reversibility* usually is toned down quite a bit, although the ideas of forgiveness, redemption, salvation, living after dying amount to a kind of *reversibility*. The miracle recounted as the resurrection of Lazarus after his body already stinketh is instructive:

> And when he [Jesus] thus had spoken, he cried with a loud voice, Lazarus, come forth. And he that was dead came forth. John 11:44-45.

That is the *reversibility* many of us used to dream about! Too bad in the real world, death is an irreversible process involving critical organ failure of one type or other. Jesus own death and resurrection supposedly was executed as a way of reversing the damage caused by Adam and Eve. We sometimes hope that by confessing to an improper act we may return to the good graces of a superior, whether it be a boss, parent, spouse, or god.

The simple act of reversing your vehicle or of returning to your home is familiar to most of us. The slogan "What goes up must come down" is a common way of stating a covert belief in *reversibility*. Any thoughts using "shoulda, woulda, couldas" consider what might have been if only we could reverse time. Infants are especially delighted to play hide-and-seek, watching the repetition of events, and then learning to predict their reoccurrence, just like a real scientist. It takes a while to discard the juvenile belief in *reversibility*. Coming from the religious tradition, many folks are unsure about whether time actually could be reversed. That is why popular science magazines find that articles entitled something like "Can time be reversed?" amount to highly profitable clickbait.

Reversibility and Reactionaries (Regressives)

Reaction takes many forms. Strictly speaking, all of us are reactionaries—we always are reacting to one event or another. Our reaction might be progressive or regressive. In politics, the word reactionary unfortunately has taken on the connotation for those who think it possible to return to a former period. I think that usage should be discarded. The proper replacement is "regressive" for those who wish to return to a former period. Conservatives like things just the way they are. Progressives want to "change the world," for the better, they think.

Religions that favor creation tend to be conservative. Earth was created just for them just the way it is. There is no reason to go forward or backward in the way conservatives live their lives. Each year is the same as the previous. But the macrocosm—the environment—is always changing. A step out of the conservative culture leads to new ideas about the external world and what can be done about it. Innovations lead to lifestyle changes that threaten the old ways. "Generation gaps" develop. The young stereotypically become progressive and the elderly remain conservative. Faced with a new, unfamiliar environment, conservatives may yearn for the "good old days," becoming regressives.

Attempts to put those yearnings into action require a belief in *reversibility*. If you did not believe things actually could be reversed, you would be unlikely to attempt to do so. For instance, evangelicals opposed to abortion probably believe that overturning Roe v. Wade will be permanent. Regressives hoping to eliminate Social Security and Medicare may believe those socialistic programs will never return.

The regressive belief in *reversibility* has much in common with the use of that assumption in systems philosophy. In the laboratory, we often considered certain chemical reactions to be completely reversible. Of course, that was not based on a fundamental assumption. It was only a specific assumption applicable to particular systems as long as we did not consider the environments in which those reactions took place. Similarly, the regressive analysis fails whenever the environment of a particular political program is ignored.

The *reversibility* assumption is common in politics, particularly where religion plays a significant part. Thus, Roe v. Wade was the product of political pressure by an expanding movement where women sought to gain freedom from biblical strictures in which they traditionally were considered the property of men.[99] In the past, there were few women in science. Now there are as many as 20% women in physics, 40% in astronomy, and 50% starting out in soil science. These "environmental" changes are not temporary. There will be no going back to the patriarchy of the "good old days." Similarly, social programs instituted in response to the Great Depression and the necessity and desire for women to be employed are not really "reversible." Attempts to do so inevitably reveal why those programs were instituted in the first place. The 2020 Pandemic recession has resulted in socialistic programs suited to that occasion. No doubt, regressives will try to reverse those programs when the recession ends.

---

[99] Tarico, 2017, Trusting doubt: A former evangelical looks at old beliefs in a new light. [Also, see her wonderful interview at: https://go.glennborchardt.com/fm-Tarico.]

Irreversibility and Science

Lab work often involves the use of experimental protocols designed to reverse previously performed reactions. As in systems philosophy in general, specialists can ignore insignificant portions of the environment in reversing or repeating an experiment and still get useful results. Of course, because the universe is infinite, none of these repetitions will produce the *exact* same result since all events are ruled by **causality**. That is why we "control" experiments, trying to avoid the infinite myriad of things that would "mess them up."

Most specialists probably do not care one way or the other about whether *reversibility* or **irreversibility** is correct. That is a philosophical question and scientists, even those called "Philosophical Doctors," usually take no courses in philosophy. I know I didn't. It was only my wide reading outside my specialty that triggered my interest in scientific philosophy and its possible application to understanding the absurd claims being made by theoretical physics and cosmogony.

Reversibility and Relativity

Relativity's compromise between religion and science was regressive, as all such compromises must be. By rejecting Newton's Second Law of Motion whenever inconvenient, physics rejected reality in favor of religious ideality. The Second Law was simple for describing the physicalness of physics: the collision of one thing with another. Giving that up was promoted as a great advance, an occasion for the appellation of the word "modern" to physics. It was anything but that; it was the epitome of *reversibility*, and a welcome mat for anti-science dreams and imaginings.

Many fantasies are derived from relativity. From time travel to wormholes, believers in the supernatural can receive daily missives from the smartest folks that would titillate those weary of their hum-drum lives in the real world. Regressives and reformists can debate some of these incessantly. Here are some of them:

Time Travel

When I get that time-worn question about going back in time, I give this answer as a sort of "proof" of *irreversibility*: Suppose you wanted to go back to a time a hundred years ago. Remember, even Einstein knew all things are in motion with respect to all other things. It turns out the night sky is unique at each particular moment. If you wanted to go back a hundred years, you would have to move all those stars and all those galaxies back to the positions they were in at that time. Good luck with that!

Even the regressives who write for Wikipedia are dubious: "It is uncertain if time travel to the past is physically possible." Nonetheless, to continue their homage to the great man, they must admit:

> As for backward time travel, it is possible to find solutions in general relativity that allow for it, such as a rotating black hole. Traveling to an arbitrary point in spacetime has a very limited support in theoretical physics, and usually is connected only with quantum mechanics or wormholes, also known as Einstein-Rosen bridges.[100]

Don't these fellows ever check the math that gives such stupid results? Is that what is supposed to prove "Einstein is always right?" Have they never looked up at the stars and thought of assuming *irreversibility*? Maybe it is time now, but then, of course, they would have to throw out the rest of relativity's indeterministic interpretations too. Perhaps Wikipedia also should throw out the entire entry on Time Travel. On the other hand, it provides a rich lesson on what is wrong in regressive physics. It includes a rogue's gallery of regressive perpetrators along with this statement:

---

[100] https://go.glennborchardt.com/timetravel

Religious Roots of Relativity

> *...but physicists cannot come to a definite judgment on the issue without a theory of quantum gravity to join quantum mechanics and general relativity into a completely unified theory.*

Sounds like that would be a completely unified mess.

Many-worlds Theory

Here is the other half of what you have to work with: Among the inanities wrought by quantum mechanics is the Many-worlds Theory. This hypothesizes many oxymoronic "universes" in which a time traveler could experience a different (or the same) history as the one they came from. Egads! Folks get *paid* for this? Just think. You have to study stuff like this to become a physicist nowadays. Even cartoonists understand the religious entanglement (Figure 9):

Figure 9. Cartoonist's take on the entanglement between quantum mechanics and religion. Credit: Mohammed Jones (https://www.jesusandmo.net/).

CHAPTER 8: FINITY

The Eighth Assumption of Religion

The universe is finite, both in the microcosmic and macrocosmic directions.

Throughout history, this assumption has had its detractors. Now, of course, it has widespread acceptance in both science and religion.

Finity and Religion

As mentioned, *creation* myths common to most religions hypothesize the *creation* of something out of nothing. Generally, those myths concentrate only on the tiny portion of the universe familiar to the theoretician. First it would be Earth, then the solar system, and then the galaxy. These were considered finite entities within a great emptiness. The logic follows from the observation that everything we know had a beginning. And all things have boundaries, some not so clear, but nevertheless boundaries. Anything with a boundary is finite.

How this *creation* could have occurred always has been a mystery. Typically, religious dreamers invented a god that was supposed to have done the job. That settled it—for millennia. The very existence of the universe was considered evidence for god or for whatever gods you imagined.

Infinity and Science

From time to time, scientists have entertained ideas of both microcosmic and macrocosmic infinity. Aristotle assumed matter and motion to be infinitely subdividable but that the universe was finite. Newton assumed the universe was infinite but that matter was finite. I finally decided that logically, there was no special reason to split *infinity* on the basis of scale. That is when I included *infinity* in my definition of univironmental determinism, the observation that what happens to a portion of the universe is determined by the *infinite* matter in motion within and without.

Except for math-heavy fields such as Artificial Intelligence, most scientific disciplines seldom are faced with having to make a choice between *finity* and *infinity*. But on close examination, it seems everything we work with is made up of other things—ad infinitum (Figure 10). The discovery of an estimated two trillion galaxies has not discouraged astronomers from looking for more—the planned Webb telescope is proof of that. Failure to find a finite "god particle" or to find the end of the finite universe are not evidence one way or the other. We will never know for sure whether the universe is finite or infinite. The best we can do is to *assume* one or the other.

Once we switch from the religious assumption of *finity* to the scientific assumption of *infinity*, there will be no turning back. That is why I call it the "Last Cosmological Revolution." Although infinity itself prevents us from declaring the ultimate end to anything, much less one of our theories, this one is different. Never again will we be able to declare a necessarily small, finite portion of the universe as an end unto itself—as a microcosm without a macrocosm. Of course, all scientific theories need modifications from time to time. Infinite Universe Theory will be no different, but its revolutionary character will never change, just as the revolutionary character of neo-Darwinian evolution will never change even though it forms only a small subset of the universal mechanism of evolution.

Figure 10. Sid Harris gets it right again: Particles within particles within particles ad infinitum. Credit: ©Sidney Harris.[101]

Finity and Relativity

Hopefully, you are getting an appreciation of the logical connection between relativity and the Big Bang Theory. You also know cosmogony, the study of the origin of the universe, contains the hidden religious assumption of *finity*. The Big Bang Theory is humanity's final *creation* myth. From Einstein's Untired Light Theory, to the misinterpretation of Hubble's cosmological redshift, to Einstein's 4-dimensional space-time, to the ad hocs such as inflation due to the magical, matterless motion called "dark energy," the Big Bang Theory is an Einsteinian mess.

There have been many objectors, as I have pointed out. Mainstream reformists have continued to use the universal expansion trope to propose

---

[101] http://sciencecartoonsplus.com/index.php

oxymoronic "multiverses," which fall into line with General Relativity Theory and the wormhole nonsense. Others, not liking the prospect of an end to the hypothesized finite universe, have proposed that the magical expansion might be followed by an equivalent magical contraction per *finity* and *reversibility*.

Olbers' Paradox

For those of you starting to believe the universe is infinite, Heinrich Olbers (1758–1840) has a question for you: "If the universe actually is infinite, then why is the sky dark at night?" Light from an infinite number of stars would keep the sky as bright as the Sun both day and night. According to the regressives writing for Wikipedia:

> The darkness of the night sky is one of the pieces of evidence for a dynamic universe, such as the Big Bang model.

By "dynamic" they really mean "expanding" or "evolving." The infinite universe is dynamic too, with everything in motion with respect to everything else. But the infinite universe cannot expand because there is nowhere for "it" to expand into. According to the Tenth Assumption of Science, **interconnection** (All things are interconnected, that is, between any two objects exist other objects that transmit matter and motion) even its miniscule portions are filled with matter. Also, it cannot evolve, for evolution is motion and the infinite universe cannot move with respect to anything else because there *is* nothing else. Only individual portions of the universe can evolve, per univironmental determinism, the universal mechanism of evolution. What happens to a portion of the universe is determined by the infinite matter in motion within and without. The infinite universe obviously has no "without."

By now you may have figured out what Obers' incorrect assumption was: Perfect transmission of light. Almost a century later, Einstein's Untired Light Theory used the same assumption (Table 3). The regressives had to follow suit, recognizing the discovery of the cosmological redshift had changed things a bit. The wavelengths from distant light sources increased, eventually becoming so long they would not appear in the visible part of the spectrum. That fact, however interpreted, clearly was the answer to Obers: The night sky is dark because of the redshift, which would occur even if the universe was infinite. Contrary to the regressive quote above, the darkness of the night sky is no evidence the universe is finite or expanding.

Rescuing Big Bang Theory

Still, the regressive persistence goes something like this: At first, the cosmological redshift was supposed to be a result of a mediumless Doppler Effect produced by galaxies receding due to universal expansion. When telescopes improved and redshifts got huge, their calculations showed greater than *c* recession. That's where things got really strange. To save Einstein's constant *c*, regressives then claimed empty space itself was magically expanding—another ad hoc to the rescue. Big Bang Theory was saved, at least for those who could stomach what came with it: the obligatory early super rapid expansion of the universe hypothesized in inflationary universe theory.[102] Big Bang Theory gets increasingly bizarre with every ad hoc used to prevent its falsification. I wonder how they will explain those elderly galaxies recently observed at distances supposedly 13 billion light years away.[103] They look like our Milky Way galaxy, which is about 13 billion years old. That light left them 13 billion years ago, so they must be at least 26 billion years old now.

This bears repeating. In Infinite Universe Theory, the cosmological redshift has a much simpler explanation, which was favored by the contrite elder Hubble: Tired light. All matter and all the motion of matter gets tired over distance, just like you do when running down the street. Like everything else, light loses energy going from point A to point B, which is what the increases in wavelength indicate. Because light is a wave, its

---

[102] Guth, 1998, The Inflationary Universe: The Quest for a New Theory of Cosmic Origins; Guth and Steinhardt, 1984, The inflationary universe.
[103] Borchardt, 2017, Infinite Universe Theory, Ch 4.

velocity is determined by the aether medium. The velocity of light stays constant over distance in so far as the properties of the medium remain constant. About the only way light loses energy is through increases in wavelength. This satisfies the Second Law of Thermodynamics and rejects Einstein's claim light exhibits perpetual motion. Waves lose energy over distance because it is impossible for any wave to produce an exact duplicate of itself. This inability appears as a lengthening of the wave. It is impossible for waves to get shorter over distance because that would require a magical energy input. Being religious, maybe the regressives can come up with an ad hoc for that too.

Glenn Borchardt

CHAPTER 9: ABSOLUTISM

The Ninth Assumption of Religion

Identities exist, that is, any two things may have identical characteristics.

The statement above is a simple test to see if you lean toward *absolutism*. *Absolutism* is much broader than that. It has much in common with the First Assumption of Religion, *immaterialism*. And, like all the other Assumptions of Religion, *absolutism* is an attempt to convert what is imagined into what is real. Folks who do that often are called idealists. They might say: If the universe is infinite, then there must be another person *exactly* like me somewhere else. This, of course, is a contradiction of microcosmic infinity. With matter being infinitely subdividable, there is no way any two things can be identical.

Absolutism and Religion

Classical idealism was perhaps best represented by Plato, who considered his imagined geometric shapes to be perfect representations of the imperfect shapes seen in the real world. Plato and his followers appeared to be unaware of the reason imperfection exists: The Infinite Universe cannot produce perfection. The "perfect" sphere and the "perfect" diamond always have imperfections—one just needs to have high enough magnification to see them.[104] If the universe really consisted of the perfect

---

[104] Reviewer Steven Bryant writes: "True in a physical sense. But…if Einstein says a shape is a sphere and it is not a mathematically perfect sphere, then it is not a sphere. If one does not demand perfect precision, then one can interpret a flat disc as being a

spheres of the atomists, we would not be here. We are the products of imperfection.

Moral Absolutism

Here is a good bit on *absolutism* in religion:

> *Many religions have morally absolutist positions, regarding their system of morality as having been set by a deity or deities. They therefore regard such a moral system as absolute, (usually) perfect, and unchangeable. Many philosophies also take a morally absolutist stance, arguing that the laws of morality are inherent in the nature of human beings, the nature of life in general, or the universe itself.*[105]

There is much here relating to all the other Assumptions of Religion. Pronouncements about morality are attempts by which religion, derived from dreams and imagination, extends *immaterialism* into the material world. Decrees about morality range from the absolute to the relative. Absolutes, being absolute, are stated without context. They are akin to systems philosophy, which tends to ignore the environment in which the system exists. Both ascribe to *finity*. Moral decrees become more relative with the context provided by the external world. That is why the absolute version survives best when religion is shielded from the outside world as in cloisters and fundamentalist sects closely adhering to a particular sacred text. The *absolutism* "Thou shalt not kill" works fine until someone starts shooting at you.

---

sphere—which is what Einstein does in his [SRT] proof." See Bryant and Borchardt, 2011, Failure of the relativistic hypercone derivation.
[105] https://psychology.wikia.org/wiki/Moral_absolutism

Political Absolutism

Sometimes the shield against the outside world does not work so well. King James I (who oversaw a new English translation of the bible) was a strong advocate of royal *absolutism*, a kind of "divine right of kings." Unfortunately, this was frowned upon by his increasingly egalitarian Parliament, leading to much verbal conflict. A rebellion occurred against his successor, son Charles I, after he made similar claims, leading to a civil war. That was followed by accusations of treason, his beheading in 1649, and the establishment of a short-lived republic. Priests and dictators are nothing without *absolutism*.

Mathematical Absolutism

Like religion itself, mathematics extends imagination into the real world via abstraction. Because everything in the real world has an infinite number of characteristics, an abstraction can never give a complete picture of reality. Thus, we give an absolute, finite, and certain quality to our claims that $1 + 1 = 2$, even though there are no examples in the real world to which that equation could apply with absolute perfection. This is where regressive physicists and cosmogonists enamored with math often get into scientific trouble. Math does not prove the real world; the real world proves math. As an example, one of the main themes of this book is that Einstein's mathematical abstraction that space is perfectly empty is an idealization and therefore cannot possibly be correct. In another example, when we say two mates are a couple, we do not mean they are identical. The math, however, says just that. It says nothing about their differences. They could be two males, two females, or a male and a female. Math has such limitations, but obviously has usefulness. We can provide two chairs for a couple without knowing their sex.

Again, *absolutism* in math is similar to the *absolutism* in religion. In both we try to provide *certainty* through imagining. And it works (to a degree). Although there are no perfect equations, they can tell us a lot about the material world in a way similar to religious absolutes that often

Relativism and Science

Relativism assumes: "All things have characteristics that make them similar to all other things as well as characteristics that make them dissimilar to all other things" (Table 1). It is the reason dreams about having another "you" in the Infinite Universe can never come true. Because the universe is infinite, each portion of it contains an infinite number of things within and without. In other words, nothing in the universe can be exactly as we might imagine it to be.

No Two Snowflakes Alike

This subject has been debated endlessly. A fellow nick-named "Snowflake Bentley" is famous for being the first to photograph a snowflake. Since then (1865) he photographed over 5,000 without ever finding two that were identical (Figure 11). That is apparently not enough for some absolutists. I once got into a Wikipedia battle with a fellow who used a finite number of characteristics to come up with something like only 10^{128} possibilities in his "proof" there really could be two snowflakes that were alike. I did not need to use math to claim no two snowflakes were alike per *relativism*. That went back and forth for a while, after which I gave up. I was slightly cheered after seeing the latest edition of the entry: "It is unlikely that any two snowflakes are alike due to the estimated 10^{19} (10 quintillion) water molecules which make up a typical snowflake." That estimate was from snow scientist Charles Knight at the National Center for Atmospheric Research in Boulder, Colorado. Still, his interviewer Mariana Gosnell, author of "Ice: The Nature, the History, and the Uses of an Astonishing Substance" remained an agnostic:

> *The way they can arrange themselves is almost infinite. So, you know, nobody can say for absolute certain, but I think experts are in agreement the likelihood of two being identical is next to impossible.*[106]

Glenn Borchardt

Note this is just another instance of folks being reluctant to make reasonable scientific assumptions in the face of an impossible task. This is a faceoff between a fundamental scientific assumption (*infinity*) and a religious one (*finity*). Only one can be correct. This snowflake example gives overwhelming evidence for *infinity*, but the Infinite Universe, being infinite, will never yield 100% proof for it. As scientists, we must go the rest of the way, *assume infinity*, and go on with our work. In this case, even tiny snowflakes are telling us the universe is infinite and that there is no chance in hell the *finity* required for Big Bang Theory is valid.

Figure 11. Some of the infinite number of possible snowflakes. Credit: <u>Snowflake Bentley Website.</u>

Matter-Space Continuum

The matter-space continuum nicely illustrates the difference between *absolutism* and *relativism*. Both perfectly solid matter and perfectly empty space are idealizations. Idealizations do not and cannot exist—they are like Plato's perfect spheres; they are purely imaginary—like religion itself. Nonetheless, in science we use idealizations all the time. It helps us understand the properties of matter and space, which exist between the nonexistent extremes. We can imagine the walls in our room being ideal, completely solid matter and the doorway being ideal, completely empty space. While that is not completely true, it does not prevent us from

---

[106] Roach, 2007, "No two snowflakes the same" likely true.

avoiding the walls and escaping through the doorway. Thus, idealizations, although always imperfect cartoons of real things or real motions of things, are necessary shorthand for understanding the universe. Because the universe is infinite, nothing we can say about an idealization can be perfectly correct, but, like the "empty space" of the doorway, an idealization often is sufficiently definitive as an aid in our attempts to manipulate the external world.

The theoretical danger for us previously religious scientists is to mistakenly believe those idealizations actually exist. I must admit that as a young grad student I fell into that trap. In clay mineralogy I studied "type minerals" considered by the sellers of such as being completely pure. I found some contaminants but, to preserve my naïve idealism, I considered them to be outside the crystal structure of those minerals. I attempted to physically remove those contaminants from the sample, expecting to leave the "pure" type mineral behind. I was wrong. The "contaminants" were within the crystal structure itself. They never could be removed without destroying the mineral. I never found a mineral that was perfectly pure. None yielded a chemical analysis exactly as represented by its theoretically ideal chemical formula. Welcome to the real world Borchardt!

We will visit this subject again in the next section. In the meantime, be sure to read more details about the matter-space continuum and similarity analysis in the Glossary.

Absolutism and Relativity

As seen from the discussion of the matter-space continuum, aether denial is based on *absolutism*. The naïve assumption space is perfectly empty pesters regressive physics and cosmogony to this day. Like perfectly solid matter, perfectly empty space has been found nowhere. No one has been able to attain absolute zero, which would indicate a complete absence of the vibrations of matter. Outer space is a virtual particle zoo. The discovery of the Cosmic Microwave Background indicates intergalactic space has a temperature of $2.7°K$. Temperature is the vibration of matter, so there has to be matter there, whether one calls it "aether," "dark matter," hydrogen, or whatever. If outer space contained the perfectly empty space of the idealist (like Einstein) its temperature would be $0.0°K$.

Empty Space and Nonexistence

Empty space and nonexistence are the same concepts. As shown above, perfectly empty space is an idealization. It cannot exist. That means nonexistence is impossible. It is impossible for the universe not to exist—everywhere and for all time. To believe otherwise is to be a naïve idealist. You have a choice: *absolutism* and *finity* or **relativism** and **infinity**. Which do you choose?

Absolutism and Untired Light Theory

The other idealisms required for Einstein's Untired Light Theory cannot exist either. There can be no massless, immaterial light particles having perpetual motion without losing the energy clearly displayed by the cosmological redshift. The Untired Light Theory used to support the expanding universe hypothesis is invalid because it is based on imaginary, impossible religious idealizations.

Absolutism and Finite Particle Theory

In the discussion of the matter-space continuum we showed that perfectly solid matter was an impossible idealization. Finite Particle Theory, which uses that idealization in violation of **relativism**, was first imagined by the Greek atomists. Solid, indivisible identical atoms were surrounded by an equally imaginary void that allowed their movement. That theory was falsified when various chemical elements were found to have varying properties. In other words, each type of element had constituents, submicrocosms similar to the ones that make the $E=mc^2$ equation work. That did not mean, of course, that the constituents of those submicrocosms could not be the finite, solid atoms hypothesized by the Greeks. They would be the ultimate constituents of matter, containing the imagined crème-pie filling, so to speak, that gave all things mass.

Particle physics has conducted a lengthy, costly campaign to discover the constituents of what we now call atoms. The primary instrument for this is the particle accelerator, with the latest model and its maintenance costing billions. Many submicrocosms have been found, but none have qualified as the sought-for finite particle filled with solid matter. The most recent candidate is a cousin of the "photon" called the Higgs boson dubbed the "god particle" by the media. Among the many problems with this particular accelerator rubble is its mass being over 125 times anything so far found inside the lightest atom. On top of that, it supposedly gives mass to an atom by acting like a sort of molasses in its surroundings. Well, you know, sort of like the mass my SUV gets when stuck in a snowbank. Then too, it doesn't last very long, decaying into four muons after 10^{-22} seconds. As might be expected for a billion-dollar result, regressives made a big deal over the Higgs discovery, reporting it internationally, and giving each other Nobels over it. At least one dissident theoretical physicist was not impressed.[107]

Per *relativism*, the finite, solid particle that fills all matter never will be found. Matter is an abstraction for all things. Per univironmental determinism, each thing contains other things and has other things outside it, ad infinitum. Although further exploration of some of those constituents may be interesting and even useful, Finite Particle Theory is a waste of time and money.

---

[107] Unzicker, 2013, The Higgs Fake: How Particle Physicists Fooled the Nobel Committee.

CHAPTER 10: DISCONNECTION

The Tenth Assumption of Religion

There may be perfectly empty space between any two objects.

Remember the atomists had to imagine a void to go along with their imagined solid atoms, which were supposed to be the constituents of all things. This last of the Ten Assumptions of Religion has much broader implications as well.

Disconnection and Religion

By assuming space is empty in tune with *absolutism*, *disconnection* is consupponible with the religious assumptions of *immaterialism* and *finity*. Each, in their own way, are like systems philosophy, which tends to ignore surroundings. Systems philosophy might equivocate, but *disconnection* does not. *Disconnection* amounts to an outright proclamation that perfectly empty space can exist.

Denigration of Matter

With religion originating from dreams and imaginings tending to favor the dreamer, the solipsist tends to ignore the external world. Strict devotion to religion requires the adherent to assume these dreams and imaginings are more important than real things. You are unlikely to convince the believer in heaven and hell that they don't exist somewhere. Because the external world consists only of matter, famous radical solipsists such as

Chopra and Sheldrake tend to denigrate matter whenever they don't deny its existence altogether. As I mentioned before, one instructive title of a Chopra book is: "How consciousness became the universe." Sheldrake is less extreme. The abstract to his "The Science Delusion" asserts that: "Sheldrake shows that the materialist ideology is moribund." He holds out hope to the devout that there is more to the universe than matter and with that belief science will be more interesting and more fun. I am afraid we already have had enough of that "fun" in the form of empty space, 4-dimensions, expanding universes, wormholes, and what have you. If you think those ideas are "fun," maybe you should be going to church every week. These radical ideas are not particularly popular among mainstream scientists, but they have a huge audience among those still holding religious assumptions. Books pushing them are sure to be best-sellers among the new-age crowd.

Old-time religion, descended from dreams and imaginings, like those of Chopra and Sheldrake, has a tendency to denigrate matter. Who has not heard of the "sins of the flesh" trope? You are taught to be disgusted by your body and ashamed of your nakedness. You are asked to "sacrifice" your body in combat, with the promise your spirit will live on forever. Your actual material existence comes second to your future in an imagined heaven. Marx was right when he famously wrote: "Religion is the sigh of the oppressed creature, the heart of a heartless world, and the soul of soulless conditions. It is the opium of the people."[108]

Biological Disconnection

Religions tend to ascribe to the "Myth of Exceptionalism,"[109] the claim humans are special and not subject to the physical constraints common to other animals. There is even a radical sect that calls themselves "Breatharians," who believe they only need air and light for sustenance.

---

[108] Marx, 1843, Critique of Hegel's 'Philosophy of Right.'
[109] Borchardt, 2007, The Scientific Worldview, Ch. 13.

Young-Earth creationists see themselves as having no connection to other primates. Per *absolutism*, the obvious similarities must not be recognized. I imagine the watchword at the zoo must be "I see nodding!" as frequently remarked by Sargent Schultz in the old TV sitcom "Hogan's Heroes."

Social Disconnection

Throughout history, the distinction between friend and foe has been critical for survival. That is why we are keenly interested in the activities of others. Like the wild herbivore necessarily checking the bushes for every sound, smell, and movement, we need to distinguish between "us" and "them." That tendency will always be with us, but its degree of intensity is contextual—it depends on scarcity. In the increasingly wealthy and civilized portions or the world, millions of us have learned to get along without much lethal conflict.[110]

Social disconnections were paramount in religious history. Religious texts describe wars against adjacent tribes probably forced to invade due to overpopulation, resource depletion, and even starvation. The us/them disconnections between various religious sects leading to us/them disconnections in war are still with us today. The evolutionary purpose of religion remains: instill and enforce loyalty. The xenophobia that sees strangers as enemies only diminishes when we find it easier to share resources instead of fighting over them—the essence of civilization. What will happen when loyalty becomes international instead of national? What will happen to war—and religion?

Interconnection and Science

Remember what ***interconnection*** states: "All things are interconnected, that is, between any two objects exist other objects that transmit matter and motion." ***Interconnection*** is perhaps the most

---

[110] Pinker, 2011, The Better Angels of our Nature.

important fundamental scientific assumption underlying the universal mechanism of evolution: univironmental determinism. Absolutists have the most difficulty understanding this. Like the atomists, they think of matter as perfectly solid bits, separated by perfectly empty voids. But as I have mentioned repeatedly, perfectly solid matter and perfectly empty space does not exist. Nonexistence is impossible, from the largest portion of the universe to the smallest.

Absolutists, assuming matter to be perfectly solid, sometimes present their argument against *interconnection* as what they call the paradox of the "block universe."[111] That idea goes something like this: If matter is infinitely subdividable, then each XYZ portion of the universe would be filled with matter. The universe then would be solid matter and nothing within would be able to move.

You can see that the incorrect assumption in this particular paradox is the opposite of the one that made Olbers' Paradox false. The block universe paradox assumes matter is perfectly solid and Olber's Paradox assumes space is perfectly empty. The reality, of course is that, per *relativism*, each subdivision of what appears to be matter or space results in what appears to be matter or space. The universe does not favor one or the other. A block universe is no more possible than an empty universe.

In agreement with Aristotle, we assume there is no end to the subdivision of matter. In our book "Universal Cycle Theory,"[112] we speculated that space was filled with aether-1 particles and that these aether particles formed from still smaller aether-2 particles, and these were formed from aether-3 particles, ad infinitum. One never gets to perfectly

---

[111] This idea was proposed to me in 2014 by William Westmiller in his critique of *interconnection*. At least it is not as idiotic as General Relativity Theory's offspring, which more commonly is known as the block universe: "According to the block universe theory, the universe is a giant block of all the things that ever happen at any time and at any place. On this view, the past, present and future all exist — and are equally real" (Wikipedia, 2020).

[112] Puetz and Borchardt, 2011, Universal Cycle Theory.

solid matter or perfectly empty space. In a generally unacknowledged 1920-speech, even Einstein once agreed space contained aether and had to have at least some limited properties.[113] Of course, he never followed up on that minor lapse and left us with perfectly empty space and plenty of einsteinisms to cuss and discuss.

Disconnection and Relativity

Special Relativity Theory

Light

Like Einstein's "Untired Light Theory," *disconnection* assumes space is perfectly empty. As discussed in the *relativism* section, that assumption makes Einstein an idealist of the first rank. But that, along with seven other ad hocs (Table 3) is what it took to make the long-distance wave motion of light into the long-distance particle motion of light.

$E=mc^2$

Einstein's erroneous interpretation of Maxwell's equation hypothesizes the miraculous conversion of mass into massless energy that flits through space like the proverbial ghost. Without aether, the process once again supposedly involves perfectly empty space. The *disconnection* means there is nothing in the environment to which the internal motion of the microcosm can be transferred.

In addition, I must mention that the $E=mc^2$ equation would not work with the aforementioned concept of ideal solid matter. Solid matter could have nothing inside of it that was in motion, also violating ***inseparability*** as well as ***relativism***. That is just another reason Finite Particle Theory is a waste of time. Improvements in instruments would discover tinier and tinier particles, but none of them would be solid matter and none of them

---

[113] Einstein, 1920, Ether and the theory of relativity.

could be considered the ultimate, undividable finite particle responsible for all mass.

General Relativity Theory

Space-time

Four-dimensional space-time might be considered the greatest *disconnection*. Like "attraction," it is a concept resulting from a kinetic equation in which the real physical cause is unknown. That makes it nice for imaginings of all types. For instance, in the LIGO experiment, space-time, assumed to be empty space, was said to alternately compress and decompress as Einstein's predicted "gravitational waves" traveled toward Earth. Of course, the measured waves were merely "shock waves" traveling through the aether at the speed of light. They had nothing to do with the physical cause of gravitation. In other words, whenever space-time is used in regressive physics and cosmogony, you can expect the unacknowledged physical effects to be those of the aether medium.

The "curved space-time" surrounding cosmic bodies also is supposed to be perfectly empty. In the case of the Eddington experiment of 1919,[114] the curvature actually was due to refraction in the Sun's atmosphere. All cosmic bodies have decelerated aether,[115] otherwise called "dark matter," which may be responsible for some circumplanetary curvature, but gives no support to *disconnection*. As with other einsteinisms, General Relativity Theory's math may be useful, while its interpretation is out of this world.

Attraction

It is quite confusing for beginning physics students and casual observers that regressives have not one, but two magical causes for gravitation: space-time and attraction. Empty space-time is said to be

---

[114] Dyson and others, 1920, A determination of the deflection of light by the sun's gravitational field, from observations made at the total eclipse of May 29, 1919.
[115] Borchardt, 2018, The physical cause of gravitation.

curved by the presence of matter; attraction occurs when matter "pulls" other matter toward itself through empty space. Like, space-time, attraction has no known physical cause. I suspect this is what Newton was thinking of when he famously claimed "Hypotheses non fingo" (Latin for I feign no hypotheses, or I contrive no hypotheses) in an essay appended to the 1713 edition of the Principia. But remember Newton's laws of motion do not include pulls. Causes are described by Newton's Second Law of Motion: Body A collides with Body B.

Indeed, in 1718 Newton proposed a push theory of gravity in his famous book "Opticks":

> *Is not this Medium much rarer within the dense Bodies of the Sun, Stars, Planets and Comets, than in the empty celestial Spaces between them? And in passing from them to great distances, doth it not grow denser and denser perpetually, and thereby cause the gravity of those great Bodies towards one another, and of their parts towards the Bodies; every Body endeavouring to go from the denser parts of the Medium towards the rarer?*
>
> *...I see no reason why the Increase of density should stop any where, and not rather be continued through all distances from the Sun to Saturn, and beyond. And though this Increase of density may at great distances be exceeding slow, yet if the elastick force of this Medium be exceeding great, it may suffice to impel Bodies from the denser parts of the Medium towards the rarer, with all that power which we call Gravity.*[116]

This was close to an actual physical cause for gravitation despite his having gotten the density backwards and having omitted the actual cause

---

[116] Newton, 1718, Opticks, p. 325.

being the relatively high distal and low proximal aether pressure. We repeat this in our books frequently because it is little known and contradicts *disconnection*, which has been favored by religious types for over three centuries. Apparently, when your assumptions are based on dreams and imaginings, it is easy to imagine magical pulls through disconnected empty space. Personally, I was never able to do that, even when I was religious.

Magnetism

Magnetism, like gravitation, is described by a kinetic theory—it too has no physical cause. And like gravitation, a cause shall never be found until aether denial is abandoned. This denial is a bit strange in view of James Clerk Maxwell's brilliant conclusion that light had to be an electromagnetic wave through a medium: the aether. This from Wikipedia:

> *Around 1862, while lecturing at King's College, Maxwell calculated that the speed of propagation of an electromagnetic field is approximately that of the speed of light. He considered this to be more than just a coincidence, and commented "We can scarcely avoid the conclusion that light consists in the transverse undulations of the same medium which is the cause of electric and magnetic phenomena."*[117]

Even with Einstein's enduring dismissal of the aether, we continue to have what are still termed electro-magnetic waves, but are left with a theoretical *disconnection*. Today, magnetic and gravitational fields are thought to be continuous nothing (relativity) or discrete nothing (quantum mechanics). There is no "there" there. Again, the math works, so why not take a bow to religion and its silly assumptions and still get paid for your efforts?

---

[117] https://en.wikipedia.org/wiki/History_of_electromagnetic_theory#Maxwell

Glenn Borchardt

CONCLUSIONS

The Ten Assumptions of Religion served the religious apparatus well, at least financially. Billions of folks continue to live their lives based on the imaginings long ago descended from dreams. As glorious as these may appear to be, they can have no possible application to the real world. There are no gods, no heaven, no hell, no living after dying. To exist, a thing must have XYZ dimensions; to occur, an event must be the motion of something. It behooves all of us to be extremely careful when crossing the line between what is real and what is only imagined. Like religion, Einstein's philosophical blunders benefited him personally, making him and his weird mathematical derivations accepted by the masses with open arms. He came along at just the right time to hinder the accelerating progress of **materialism**, which in the 19th Century threatened to destroy religion and the political systems founded upon it.

Einstein was an agnostic—neither a theist, nor an atheist. As such, like many physicists, he was inconsistent in his use of fundamental assumptions. Like most scientists, he probably didn't know he had any. He seems not to have had a grand plan to merge religion and science in the way he did. Nonetheless, he was a genius at it. His many einsteinisms (right for the wrong reason) attest to his awareness of portions and activities of the universe that needed explaining. His usurpation and promotion of Maxwell's $E=mc^2$ equation as his own involved an improper interpretation, but that did not make the equation itself incorrect. Any failure to use the Lorentz Correction Factor properly could be used as evidence for his absurd "time dilation." His prediction that empty space nevertheless could be curved was sheer genius—refraction within what was really nonempty space would "prove him right"—for the wrong reason.

Again, one theme runs through all of relativity: The belief in perfectly empty space, which at the same time is a belief in the possibility of nonexistence. The empty space hypothesis made Einstein the most reactionary scientist who ever lived. Aside from its outright negation of *materialism* and the other assumptions of science, it was a negation of existence itself. For almost anyone except Deepak Chopra the existence of the universe is obvious. The nonexistence ad hoc necessary for Einstein's conversion of aetherial wave motion into aether-free particle motion was the first of a chain of events leading up to the Big Bang Theory. To question the imagined idealization of empty space is to question all Ten Assumptions of Religion, relativity, and the explosion of the entire universe from nothing.

In religious belief the *creation* of something from nothing is not a problem—the imaginative always can invent another god to create the god that did it, ad infinitum. But it still remains a contradiction for regressives when asked about what came before the Big Bang. I remember attending Stephen Hawking's lecture at U.C. Berkeley in 2007. As expected, he produced the old cosmogonical trope that the universe came from nothing (or an imaginary "singularity"). Hawking did think we would have enough time eventually to figure it out, reciting this quote from Woody Allen: "Eternity is a very long time, especially towards the end." In 2012, former Professor Lawrence Krauss thought he had the answer when he wrote the highly popular book, "A universe from nothing: Why there is something rather than nothing." It didn't take an astrophysicist to point out how silly the whole thing was. Rock and roll star Rick Doogie's review is a hoot!

As mentioned previously, in "The Structure of Scientific Revolutions" physicist-historian Thomas Kuhn invented the word "paradigm" for any set of theories used to advance a particular line of thought. Each paradigm becomes the victor in a difficult struggle to replace previous ways of thought. That is correct. But unfortunately, Kuhn went on to claim that scientific theories merely were results of extent sociological conditions. His first edition was published in 1962, about the time when postmodernism was appearing as a reaction to modernism. Though defective in many ways, postmodernism, like Kuhn, called much-needed attention to the macrocosm, the environment in which humans operated.

As mentioned previously, my discovery of univironmental determinism was influenced by the 1970 mass movement that started to recognize the importance of the environment. Kuhn might be regarded as the first postmodernist to analyze the history of science. He was criticized heavily for implying science was not progressive, that the "truths" uncovered by science were fad-like and not trending toward increases in understanding the fundamental character of the universe. Believers in what they called "modern" physics were not happy. The regression set in motion by Einstein was considered by them as a great advance on Newton's materialistic mechanics. That was partially true in that the regression ultimately implied, but never realized there was something wrong with the assumption of *finity*, which had been the hallmark of mechanism.

Kuhn's stance was a bit surprising in that he did not criticize modern physics in particular as we do today. Like many of us, he may have had only a vague feeling something was not right in the theoretical physics kitchen. As a physicist and agnostic, he would not be expected to realize what the regression meant and its indispensable connection to religion. It is doubtful he would have considered a return to ***materialism*** to be progressive. So, he ended up with a kind of cyclic theory of the history of science. This, of course, flew in the face of everything we know about scientific advancement. Continuing examination of the universe both microcosmically and macrocosmically invariably produces increases in data and global knowledge. Eventually, that forces paradigm shifts toward a more objective and more accurate understanding of the infinite universe. The universe cares not a whit about whether our theories are correct or not. Kuhn is right that our subjectivity tends to obscure the objects right in front of us. This is particularly true when that subjectivity arises from ancestral dreams and imaginings common to religious tradition.

That this regression has survived for over a century seems to have flabbergasted more than 9,500 dissidents trying to rid the world of its insults to logical thinking (Figure 12). But remember the religious world has not changed that much during the regression. As of 2020, 84% still are religious. Many, if not most, of the dissidents were raised in the religious tradition even though, like Einstein, they may claim to be agnostic or atheist. Also, like Einstein, many of them have no problem with suppositions unconsciously derived from those traditions or from colleagues so affected. The belief in massless particles, empty space, and matterless motion persists even though the explosion of the universe out of nothing may be suspect.

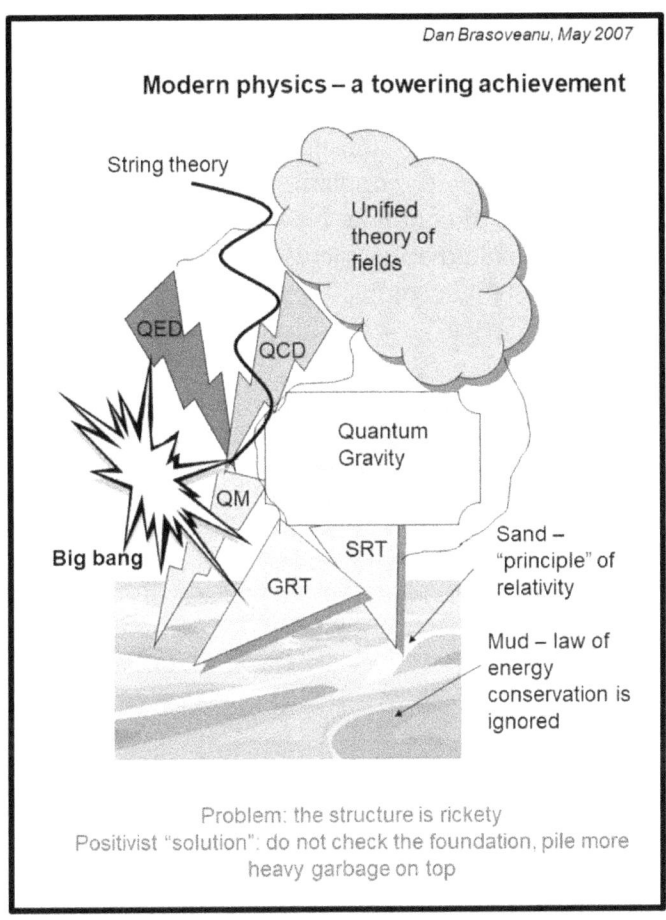

Figure 12. One reformist's view of regressive physics and cosmogony.[118]

So, what is the hope for the future? When will we ever get our feet back on the ground? When will we stop taking our cues from imaginings rooted in the dreams of our ancestors? The unaffiliated (those who profess

---

[118] Brasoveanu, 2008, Modern Mythology and Science: Crysis in Modern Physics.

no religious belief) are the largest philosophical group after Christianity and Islam. There are over 1.2 billion of us in the world. In the US the unaffiliated have increased by 30 million during the last decade alone (Figure 13) with most of this growth occurring in the younger, better-educated generations (Figure 14). While this may be hopeful, few of these Nones are well educated enough or courageous enough to admit they are atheists. Despite the relatively rapid growth of the unaffiliated, the persecution of atheists continues throughout the world. The establishment of theocracies and the rapid population growth in religious countries assures the oppression will continue. Nevertheless, the Internet age allows the curious to explore alternative views, often without penalty. Borders between countries are becoming increasingly less significant as globalization proceeds apace.

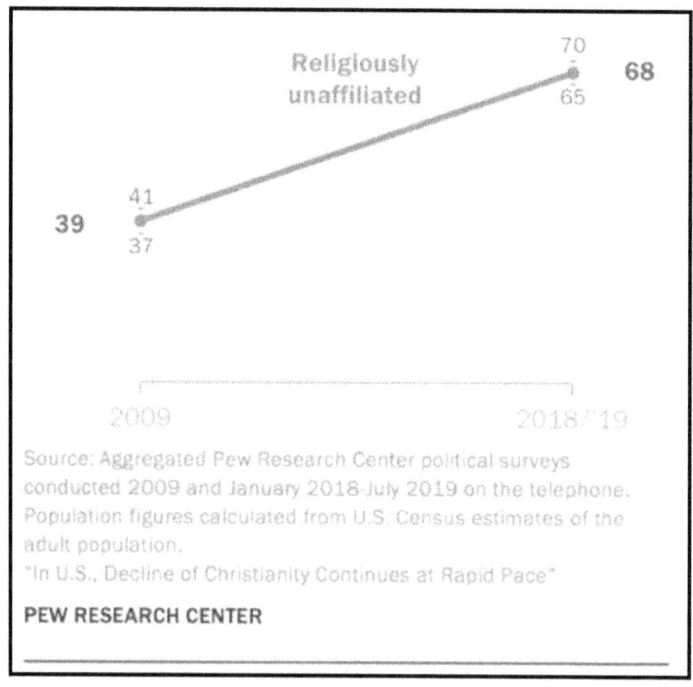

Figure 13. The 30-million increase in the religiously unaffiliated during the last decade in the US. There were 39 million in 2009 and over 68 million in 2019.

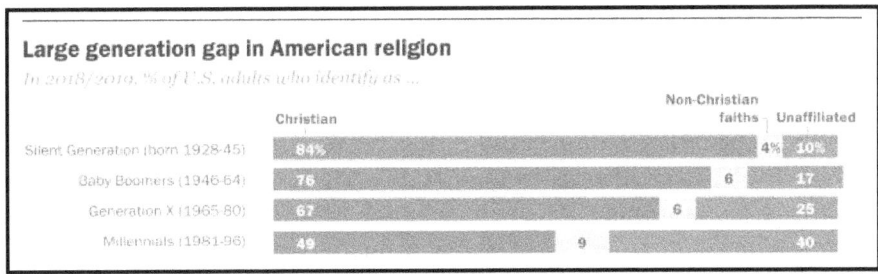

Figure 14. Religious rejection in relation to youthfulness of generations in the US. Credit: Pew Research Center.[119]

In "Infinite Universe Theory" my Chapter 18.2 "Paradigmatic Persistence and Requiem for the Big Bang Theory" summed up my prognosis. It was primarily based upon the continuing declines in global birth rates and associated declines in global economic growth rates. That analysis assumed human population growth, like that of all animals, follows a sine curve (Figure 15). And, like other animals, humans adhere to the "Principle of Least Effort."[120] The relatively easy exploitation of the world's natural resources that produced the exponential growth before 1989 has become ever-more difficult since. As our population slowly increases from 7.64 billion to our carrying capacity of 10 billion, great cultural changes will occur. The switch from growth to sustainability and from rural to urban will force philosophical changes leading to intense questioning of authority and rejection of traditional values. That will not occur without a great deal of global stress. The 2020 Pandemic Recession is only the beginning. Philosophies that originated in the dreams and imaginings of our ancestors will fall to the wayside, becoming mere curiosities in the dustbin of history. Nothing fails like prayer.[121] When the

---

[119] Pew Research Center, In U.S., decline of Christianity continues at rapid pace.
[120] Zipf, 1949, Human Behavior and the Principle of Least Effort.
[121] Benson and others, 2006, Study of the therapeutic effects of intercessory prayer (STEP) in cardiac bypass patients; Masters and others, 2006, Are there demonstrable effects of distant intercessory prayer? [This paper analysed 14 similar papers. All had null results. They recommended "that further resources not be allocated to this

biggest cultural change occurs around 2050, relativity and cosmogony will be in tatters. Theories requiring perfectly empty space and the nonexistence it implies will be forgotten. With the switch to Infinite Universe Theory, this Last Cosmological Revolution forces all of humanity to the realization we are only a fortunate, ephemeral portion of a universe that exists everywhere and for all time.

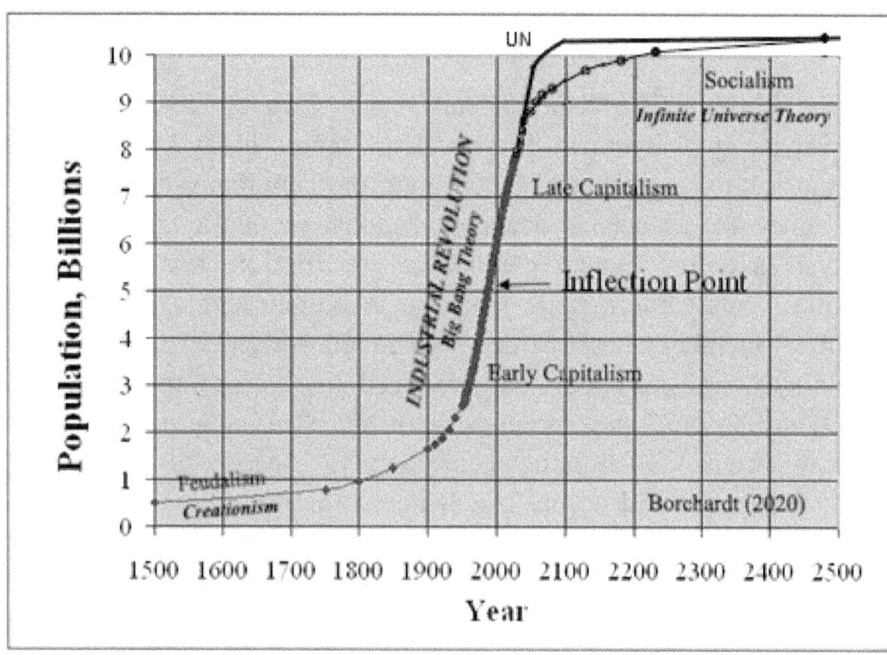

Figure 15. Sigmoidal growth curve for global population assuming perfect symmetry about the 1989 Inflection Point, including the especially abrupt change predicted by the UN in 2024 (modified from Borchardt, 2007).

line of research."]

Glenn Borchardt

References

Abu-Bakr, Mohammed, 2007, The End of Pseudo-Science: Essays Refuting False Scientific Theories Taught in Schools, Colleges, and Universities, iUniverse, 86 p.
Angel, R.B., 1980, Relativity: The theory and its Philosophy: New York, Pergamon, 259 p.
Armstrong, Dave, 2017, Albert Einstein's "Cosmic Religion": In his own words, Patheos, Accessed 20200501 [https://go.glennborchardt.com/Einstein-quotes].
Barr, S.M., 2003, Modern Physics and Ancient Faith: Notre Dame, IN, University of Notre Dame, 328 p.
Benson, Herbert and others, 2006, Study of the therapeutic effects of intercessory prayer (STEP) in cardiac bypass patients: A multicenter randomized trial of uncertainty and certainty of receiving intercessory prayer: American Heart Journal, v. 151, no. 4, p. 934-942. [https://go.glennborchardt.com/prayertest].
Borchardt, Glenn, 1974, The SIMAN coefficient for similarity analysis: Classification Society Bulletin, v. 3, no. 2, p. 2-8. [http://bit.ly/SIMAN1974].
Borchardt, Glenn, 1984, The Scientific Worldview [review manuscript]: Berkeley, California, Progressive Science Institute, 343 p. [http://doi.org/10.13140/RG.2.2.16123.52006].
Borchardt, Glenn, 2004, The Ten Assumptions of Science: Toward a New Scientific Worldview: Lincoln, NE, iUniverse, 125 p. [https://go.glennborchardt.com/TTAOSfree].
Borchardt, Glenn, 2007, The Scientific Worldview: Beyond Newton and Einstein: Lincoln, NE, iUniverse, 411 p. [https://go.glennborchardt.com/TSW].
Borchardt, Glenn, 2008, Resolution of the SLT-order paradox, in Proceedings of the Natural Philosophy Alliance, Albuquerque, NM [http://doi.org/10.13140/RG.2.1.1413.7768].
Borchardt, Glenn, 2009, The physical meaning of E=mc^2, in Proceedings of the 16th Conference of the Natural Philosophy Alliance, Storrs, CT, p. 27-31. [http://doi.org/10.13140/RG.2.1.2387.4643].
Borchardt, Glenn, 2011, Einstein's most important philosophical error, in Volk, Greg, ed., in Proceedings of the 18th Conference of the Natural Philosophy Alliance, College Park, MD, Natural Philosophy Alliance, Mt. Airy, MD, p. 64-68. [http://doi.org/10.13140/RG.2.1.3436.0407].
Borchardt, Glenn, 2017, Infinite Universe Theory: Berkeley, California, Progressive Science Institute, 337 p. [http://go.glennborchardt.com/IUTebook].
Borchardt, Glenn, 2018, The physical cause of gravitation: Preprint. [http://vixra.org/abs/1806.0165].
Borchardt, Glenn, Aruscavage, P.J., and Millard, H.T., 1972, Correlation of the Bishop Ash, a Pleistocene marker bed, using instrumental neutron activation analysis: Journal of Sedimentary Research, v. 42, no. 2, p. 301-306. [https://go.glennborchardt.com/BAM72].
Brasoveanu, Dan, 2008, Modern Mythology and Science: Crysis in Modern Physics, iUniverse, 94 p.
Brewis, Roger, 2013, Physics, rogue science?: Alternative theorists

[https://go.glennborchardt.com/de-Climont-group].
Bryant, Steven 2011, The twin paradox: Why it is required by relativity, in Volk, Greg, ed., in Proceedings of the 18th Conference of the Natural Philosophy Alliance, College Park, MD, Natural Philosophy Alliance, Mt. Airy, MD [http://www.relativitychallenge.com/papers/].
Bryant, Steven, and Borchardt, Glenn, 2011, Failure of the relativistic hypercone derivation, in Volk, Greg, ed., in Proceedings of the 18th Conference of the Natural Philosophy Alliance, College Park, MD, Natural Philosophy Alliance, Mt. Airy, MD, p. 99-101. [http://doi.org/10.13140/RG.2.1.1404.8406].
Bryant, S.B., 2016, Disruptive: Rewriting the Rules of Physics: El Cerrito, California, Infinite Circle Publishing, 312 p. [http://go.glennborchardt.com/Bryant16]
Bryant, S.B., 2019, An open response to Johanna Miller's column: 'Sorry, crackpots', http://stevenbbryant.com/, Accessed 20200818 [https://go.glennborchardt.com/Bryant19].
Chopra, Deepak, 2009, Quantum Healing: Exploring the Frontiers of Mind/Body Medicine NY, Random House, p. 71-72, 74.
Chopra, Deepak, 2015, How Consciousness Became the Universe: Quantum Physics, Cosmology, Evolution, Neuroscience, Parallel Universes: Cambridge, MA, Cosmology Science Publishers, 574 p.
Chopra, Deepak, and Kafatos, M.C., 2017, You are the Universe: Discovering Your Cosmic Self and Why it Matters, Harmony, 290 p.
Coe, Lee, 1969, The nature of time: American Journal of Physics, v. 37, p. 810-815. [https://go.glennborchardt.com/Coe69].
Collingwood, R.G., 1940, An Essay on Metaphysics: Oxford, Clarendon Press, 354 p.
Coyne, J.A., 2015, Faith Versus Fact: Why Science and Religion are Incompatible, Viking, 336 p.
Dadkhah, Mehdi, and Borchardt, Glenn, 2016, Guidelines for selecting journals that avoid fraudulent practices in scholarly publishing: Iranian Journal of Management Studies, v. 9, no. 3, p. 529-538. [https://go.glennborchardt.com/Guidelines-2016].
Dadkhah, Mehdi, Borchardt, Glenn, and Maliszewski, Tomasz, 2016, Fraud in academic publishing: Researchers under cyber-attacks: The American Journal of Medicine, v. 130, p. 27-30. [http://doi.org/10.1016/j.amjmed.2016.08.030].
de Climont, Jean, 2012, The Worldwide List of Dissident Scientists, 1700 p.
de Climont, Jean, 2020, The Worldwide List of Alternative Theories and Critics, Editions d' Assailly, 2679 p. [http://go.glennborchardt.com/declimont16dissidentlist].
de Sitter, Willem, 1913, An astronomical proof for the constancy of the speed of light (English translation): Physikalische Zeitschrift, v. 14, p. 429. [http://go.glennborchardt.com/desitter13light].
Descartes, Rene, 1644 [1991], Principles of Philosophy: Boston, MA, Kluwer Academic, 324 p. [http://go.glennborchardt.com/Descartes1644].
Dyson, F.W., Eddington, A.S., and Davidson, C., 1920, A determination of the deflection of light by the sun's gravitational field, from observations made at the total eclipse of May 29, 1919: Philosophical Transactions of the Royal Society A: Mathematical, Physical and Engineering Sciences, v. 220, no. 571-581, p. 291-333. [http://doi.org/10.1098/rsta.1920.0009].

Einstein, Albert, 1905 [1923], On the electrodynamics of moving bodies, in Einstein, A., Lorentz, H.A., Weyl, H., and Minkowski, H., eds., The principle of relativity: New York, Dover, p. 37-65. [http://www.fourmilab.ch/etexts/einstein/specrel/www/].

Einstein, Albert, 1916 [1923], The foundation of the general theory of relativity, in Einstein, A., Lorentz, H.A., Weyl, H., and Minkowski, H., eds., The Principle of Relativity: New York, Dover, p. 146-195.

Einstein, Albert, 1917 [1923], Cosmological considerations of the general theory of relativity, in Einstein, A., Lorentz, H.A., Weyl, H., and Minkowski, H., eds., The principle of relativity: New York, Dover, p. 421-432. [https://einsteinpapers.press.princeton.edu/vol6-trans/444].

Einstein, Albert, 1920, Ether and the theory of relativity, in Address given on May 5th, University of Leyden [http://go.glennborchardt.com/Einstein20recantation].

Einstein, Albert, 1926 [2004], Letter to Max Born on quantum mechanics, December 4, in Born, Max, ed., Born-Einstein Letters 1916-1955: Friendship, Politics and Physics in Uncertain Times: New York, NY, Palgrave Macmillan, p. 88.

Einstein, A., 1940, Science and religion: Nature, v. 146, no. 3706, p. 605-607. [https://go.glennborchardt.com/Einstein-lame].

Einstein, Albert, 1950, On the generalized theory of gravitation: Scientific American, v. 182, no. 4, p. 13-17. [http://www.jstor.org/stable/24967425].

Einstein, Albert, 1952 [2015], Letters to Solovine, 1906-1955: New York, NY, Philosophical Library/Open Road, p. 132-133. [http://inters.org/Einstein-Letter-Solovine].

Einstein, Albert, 1954a, Letter to Mr Gutkind impugning religion, scripture, and the idea of god, Accessed 20200501 [https://go.glennborchardt.com/AEs-atheist-letter].

Einstein, Albert, 1954b, Ideas and Opinions: New York, Crown, 377 p.

Farmer, B.L., 1997, Universe Alternatives: Emerging Concepts of Size, Age, Structure and Behavior (2nd ed.): El Paso, TX, Billy L. Farmer, 129 p.

Galaev, Y.M., 2002, The measuring of ether-drift velocity and kinematic ether viscosity within optical waves band (English translation): Spacetime & Substance, v. 3, no. 5, p. 207-224. [http://spacetime.narod.ru/5-15-2002.html].

Gardner, Martin, 1962, Relativity for the Million: New York, Macmillan, 182 p.

Gary, Stuart, 2011, Astronomers find old heads in a young crowd, Accessed 20171022 [http://go.glennborchardt.com/Gary11elderlygalaxies].

Gisin, Nicolas, 2020, Mathematical languages shape our understanding of time in physics: Nature Physics. [http://doi.org/10.1038/s41567-019-0748-5].

Gould, S.J., 1997, Nonoverlapping magisteria: Natural History, v. 106, p. 16-22. [https://go.glennborchardt.com/NOMA].

Gross, Neil, and Simmons, Solon, 2009, The religiosity of American college and university professors: Sociology of Religion, v. 70, no. 2, p. 101-129. [http://doi.org/10.1093/socrel/srp026].

Guth, A.H., 1998, The Inflationary Universe: The Quest for a New Theory of Cosmic Origins, Basic Books, 384 p.

Guth, A.H., and Steinhardt, P.J., 1984, The inflationary universe: Scientific American, v. 250, no. 5, p. 116-128, 154.

Hassani, Sadri, 2019, Massless is not nonmaterial: Skeptical Inquirer, v. 43, no. 4, p. 14-

16. [https://skepticalinquirer.org/2019/07/massless-is-not-nonmaterial/].
Hooper, W.G., 1903, Aether and Gravitation: London, Chapman and Hall, LTD., 358 p. [http://go.glennborchardt.com/Hooperpdf].
Houlgate, Stephen, Ed., 1998, The Hegel Reader: Malden, MA, Blackwell Publishing, 568 p.
Howells, William, 1949, The Heathens: Primitive Man and His Religions: London, Victor Gallancz Ltd, 268 p. [https://go.glennborchardt.com/Howells49].
Jammer, Max, 1999, Einstein and Religion: Physics and Theology: Princeton, Princeton University Press, 267 p.
Jung, C.G., 1960, Synchronicity: An Acausal Connecting Principle: Princeton, NJ, Princeton UP, 158 p. [https://go.glennborchardt.com/Jung].
Krauss, L.M., 2013, A Universe from Nothing: Why There is Something Rather Than Nothing: New York, Free Press, 240 p.
Kuhn, T.S., 1962, The Structure of Scientific Revolutions: Chicago, The University of Chicago Press, 210 p.
Kuhn, T.S., 1996, The Structure of Scientific Revolutions (3rd ed.): Chicago, University of Chicago Press, 212 p.
Lambert, Dominique, 2020, Einstein and Lemaître: Two friends, two cosmologies… [http://inters.org/einstein-lemaitre].
Larson, E.J., and Witham, Larry, 1998, Leading scientists still reject God: Nature, v. 394, no. 6691, p. 313. [http://doi.org/10.1038/28478].
Laszlo, Ervin, 1972, Introduction to Systems Philosophy: Toward a New Paradigm of Contemporary Thought: New York, Gordon And Breach, 328 p.
Lemaître, Georges, 1950, The Primeval Atom: An Essay on Cosmogony: New York, D. Van Nostrand, 186 p.
Lenin, V.I., 1908 [1927], Materialism and Empirio-Criticism: Critical Comments on a Reactionary Philosophy: New York, International, 397 p. [https://go.glennborchardt.com/lenin08].
Lewis, G.N., 1926, The conservation of photons: Nature, v. 118, no. 2981, p. 874-875. [http://www.nobeliefs.com/photon.htm].
Masters, K.S., Spielmans, G.I., and Goodson, J.T., 2006, Are there demonstrable effects of distant intercessory prayer? A meta-analytic review: Annals of Behavioral Medicine, v. 32, no. 1, p. 21-26. [http://doi.org/10.1207/s15324796abm3201_3].
Marx, Karl, 1843 [1977], Critique of Hegel's 'Philosophy of Right': Cambridge, United Kingdom, Cambridge University Press, 224 p. [https://go.glennborchardt.com/opium-of-people].
McKenna, Josephine, 2014, Pope says evolution, Big Bang are real: USA Today, October 28. [https://go.glennborchardt.com/Pope].
Michelson, A.A., and Morley, E.W., 1887, On the relative motion of the earth and the luminiferous ether: American Journal of Science, v. 39, p. 333-345. [http://www.anti-relativity.com/MM_Paper.pdf].
Miller, J.L., 2019, Column: Sorry, crackpots--A Physics Today editor explains why we're never going to publish your cockamamie theories: Physics Today, February 1, p. 1-7. [http://doi.org/10.1063/PT.6.4.20190201b].
Misner, C.W., 1978, The immaterial constituents of physical objects, in Symposium on

the Impact of Modern Scientific Ideas on Society (Einstein Centenary/UNESCO), Ulm, Germany, p. 1-4. [https://go.glennborchardt.com/Misner-78].

Moody, Richard, Jr., 2009, The eclipse data from 1919: The greatest hoax in 20th century science, in Proceedings of the 16th Conference of the Natural Philosophy Alliance, Storrs, CT, p. 1-26.

Newton, Isaac, 1718, Opticks or, a Treatise of the Reflections, Refractions, Inflections and Colours of Light (2nd ed.): London, Royal Society, 382 p. [http://go.glennborchardt.com/Newton1718Optics].

Pazukhin, Evgueny, 1997, The Christian materialism of Blessed Josemaría Escrivá: Opus Dei. [https://www.escriva.it/Ing/19970301.htm].

Perakh, Mark, 2007, Non sequitur in five parts: Professor Barr's effort to bolster his faith via modern physics and Gödel's theorem, Accessed 20200501 [https://go.glennborchardt.com/Barr03-review].

Petkov, Vesselin, 1989, Simultaneity, conventionality and existence: The British Journal for the Philosophy of Science, v. 40, no. 1, p. 69-76. [www.jstor.org/stable/687464].

Pew Research Center, 2019, In U.S., decline of Christianity continues at rapid pace, Accessed 20200501 [https://go.glennborchardt.com/PEW-19-on-US-nones].

Pinker, Steven, 2011, The Better Angels of Our Nature: Why Violence Has Declined: New York, Viking [http://stevenpinker.com/publications/better-angels-our-nature].

Popper, K.R., 2002, The Logic of Scientific Discovery (15th ed.): New York, Routledge, 544 p.

Puetz, S.J., and Borchardt, Glenn, 2011, Universal Cycle Theory: Neomechanics of the Hierarchically Infinite Universe: Denver, Outskirts Press, 626 p. [https://go.glennborchardt.com/UCT].

Rabinowitz, Avi, 2006 Acausality: The root of "true free will" & of universal emergence into existence, Accessed 20200501 [https://go.glennborchardt.com/acausality].

Rabinowitz, Avi, 2006 Mindless materialists, Accessed 20200501 [https://go.glennborchardt.com/mindless].

Rassin, Eric, Merckelbach, Harald, and Spaan, Victor, 2001, When dreams become a royal road to confusion: Realistic dreams, dissociation, and fantasy proneness: The Journal of Nervous and Mental Disease, v. 189, no. 7, p. 478-481. [https://go.glennborchardt.com/Dream-confusion].

Ricker, H.H., 2015, The origin of the equation E=mc^2, Accessed 20200501 [http://go.glennborchardt.com/Ricker15mc2origin].

Roach, John, 2007, "No two snowflakes the same" likely true, research reveals: National Geographic News. [http://go.glennborchardt.com/Roach07snowflakes].

Sagnac, Georges, 1913a, The demonstration of the luminiferous aether by an interferometer in uniform rotation: Comptes Rendus, v. 157, p. 708–710.

Sagnac, Georges, 1913b, On the proof of the reality of the luminiferous aether by the experiment with a rotating interferometer: Comptes Rendus, v. 157, p. 1410–1413.

Schlafly, R.S., 2011, How Einstein Ruined Physics: Motion, Symmetry, and Revolution in Science: Lexington, KY, CreateSpace Independent Publishing Platform, 350 p.

Sears, F.W., and Zemansky, M.W., 1960, College Physics (3rd ed.): Reading, MA, Addison-Wesley, 1024 p.

See, T.J.J., 1925, Researches in Non-Euclidian Geometry and the Theory of Relativity: A

Systematic Study of Twenty Fallacies in the Geometry of Riemann, Including the So-called Curvature of Space and Radius of World Curvature, and of Eighty Errors in the Physical Theories of Einstein and Eddington, Showing the Complete Collapse of the Theory of Relativity: Mare Island, California, United States Naval Observatory, 220 p. [https://books.google.com/books?id=55eHrgEACAAJ].

Sheldrake, Rupert, 2012, The Science Delusion, Hodder General Publishing Division [http://books.google.com/books?id=8zH9tgAACAAJ].

Shook, John, 2010, Religious certainty is a dangerous weapon, Center for Inquiry, Accessed 20200501 [https://go.glennborchardt.com/certainty].

Stirrat, Michael, and Cornwell, R.E., 2013, Eminent scientists reject the supernatural: a survey of the Fellows of the Royal Society: Evolution: Education and Outreach, v. 6, no. 1, p. 1-33. [http://doi.org/10.1186/1936-6434-6-33].

Gary, Stuart, 2011, Astronomers find old heads in a young crowd, Accessed 20171022 [http://go.glennborchardt.com/Gary11elderlygalaxies].

Tarico, Valerie, 2017, Trusting doubt: A former evangelical looks at old beliefs in a new light (2nd ed.), Oracle Institute Press, 304 p. [https://go.glennborchardt.com/fm-Taricobk].

Unzicker, Alexander, 2013, The Higgs Fake: How Particle Physicists Fooled the Nobel Committee, CreateSpace Independent Publishing Platform, 152 p. [https://go.glennborchardt.com/Higgs].

Vergano, Dan, 2014, Hubble Reveals Universe's Oldest Galaxies, Accessed 20171022 [http://go.glennborchardt.com/Vergano14elderlygalaxies].

Viereck, G.S., 1929, What life means to Einstein: The Saturday Evening Post, October 26, p. 17, 110-117.

Wall, Mike, 2020, Rare monster galaxy grew rapidly 12 billion years ago ... then suddenly died [https://go.glennborchardt.com/BGC-12Ga].

Walbridge, L.S., ed., 2001, The Most Learned of the Shia: Oxford, Oxford University Press, 277 p.

Zeigler, David, 2020, Religious belief from dreams?: Skeptical Inquirer, v. 44, no. 1, p. 51-54. [https://go.glennborchardt.com/god-dreams].

Zipf, G.K., 1949, Human Behavior and the Principle of Least Effort: An Introduction to Human Ecology: Cambridge, MA, Addison-Wesley, 573 p.

Glossary

AD HOC. "a word that originally comes from Latin and means "for this" or "for this situation." In current American English it is used to describe something that has been formed or used for a special and immediate purpose, without previous planning." (Webster) In science, ad hocs are add-ons used to save theories from falsification via special pleading.

AETHER. The proper spelling for the medium theoretically responsible for light transmission, gravitation, and the formation of baryonic matter. Its tiny particle size precludes direct observation, although its effects are well known. This spelling is from 14th century Latin. It is sometimes spelled in the 12th century Old French form as "ether." Unfortunately, I am guilty of using the "ether" spelling in other books and papers. I switched to "aether" for two reasons: 1) it avoids confusion with the organic compound and 2) its Latin form and its definition as the precursor to ordinary matter apparently has precedence with René Descartes (1596-1650) where it occurs on p. 140 of the Latin version of his 1644 "Principles of Philosophy." For more detail and speculation on this subject, see Chapter 16.2 in "Infinite Universe Theory."

BARYONIC MATTER. Ordinary matter, which includes electrons, positrons, atoms, and molecules that comprise the things of everyday existence. In general, baryonic matter can be sensed by the five senses directly or indirectly through instruments, as opposed to aether, which is matter too, but is much smaller and not so easily sensed.

CARRYING CAPACITY. In biology, carrying capacity generally occurs when a species accommodates to available resources. As carrying capacity is reached, it becomes increasingly difficult to obtain the necessities for life. For instance, if your life depended only on your obtaining gold from now-deficient streams, you probably would not last long. Homo sapiens has not yet reached carrying capacity, although there are many isolated instances where tribes and nations have reached population levels that were difficult to sustain without war and migration.

COLLIDEE. The object hit by a collider.

CONSUPPONIBLE. A description of multiple suppositions that are assumed to be true and without significant contradiction. The word appears to have been invented by R.G. Collingwood, a historian, who devised a method to discover fundamental assumptions.[122] The concept of consupponibility seems to give folks trouble. Maybe it is because they are accustomed to scores of contradictory ideas, such as walking on water, virgin birth, and living after dying. Modern physicists, in particular, are accustomed to wave-particle duality, the twin paradox, massless particles, and the explosion of the universe out of nothing. It is a mess. These serve as justification for indeterministic claims "the world is too complicated and too contradictory for us ordinary mortals to understand." An example of consupponibility would be these two (non-fundamental) assumptions: 1) that you have height and 2) that you have weight. These statements do not contradict one another. They are consupponible.

COPENHAGEN INTERPRETATION. The regressive interpretation of quantum mechanics based on *certainty*, which avoids ***infinity*** and ***causality*** by considering probability to be a singular cause. In progressive physics probability is used as a mathematical description of the infinite multitude of causes not determined in a particular experiment per ***uncertainty***.

COSMOGONIST. One who assumes the universe had an origin. See the details below:

COSMOGONY. The study of the origin of the universe. Creation myths, like the one in Genesis and in Big Bang Theory, have billions of supporters who remain steady in their beliefs. According to Google Books, the term became increasingly popular after 1710, with its popularity reaching a maximum in the 1880's and declining ever since, with only a few peaks in 1928 and 1958. Today, the term is almost never used in cosmology, probably because it is an overt admission of assumptive bias, although that was not a problem for the priest who invented the Big Bang Theory.[123]

---

[122] Collingwood, 1940, An essay on metaphysics.
[123] Lemaître, 1950, The primeval atom: An essay on cosmogony.

The theory got popular for the same reason that relativity did: appeal to indeterminists. In the battle with science, the Big Bang Theory represents the last compromise, a final stand for cosmogony. The timing is a bit off for the extremely conservative, but even the Pope now accepts it.[124] The mostly religious representatives in Congress would rather designate dollars for studies close to traditional cosmogony than to those opposed. As with relativity and most any theory or political action, understanding improves by using the old cliché: Follow the money.

DETERMINISM. The assumption that there are material causes for all effects. The philosophical foundation for the legal system and most scientific disciplines other than physics and cosmogony. It is opposed to the idea of free will, which assumes that some effects are uncaused.

DETERMINISM-INDETERMINISM PHILOSOPHICAL STRUGGLE. The interminable conflict between the two primary ways of viewing the universe. The conflict begins at birth, getting in full swing as infants begin to learn that the universe consists of matter in motion. Eventually, they learn that most of those things and their motions occur independently of them or their thoughts about them. To survive, we need to learn about how the universe works. We must manipulate our environment to obtain food, shelter, and clothing. We rearrange the material things around us, discovering that there are material causes for effects—we tend to become "determinists." This learning process continues throughout life with varying degrees of success. Often, we cannot determine the material cause for a particular effect—who can see air or aether? Without an abiding faith that there always must be a material cause nevertheless, we might have a tendency to become "indeterminists." Even so, those who gain wide experience in successfully manipulating their environments, such as scientists, engineers, farmers, construction workers, and those in other hands-on occupations may have a tendency to generalize what they have learned. They might optimistically conclude that "there are material causes for all effects," becoming radical determinists and budding scientists. Others might pessimistically conclude just the opposite; becoming radical

---

[124] McKenna, 2014, Pope says evolution, Big Bang are real.

indeterminists in spite of whatever interactions they have had with the real world.

Science tends to favor determinism and religion tends to favor indeterminism. One part of the struggle involves denial that there is even a struggle. A complete, final proof of either position is impossible even though there is much data in support of determinism and none in support of indeterminism. That is because there is no way that the causes for all effects could ever be determined. Failure to find a cause for a particular effect allows indeterminists to assume that there isn't one. Of course, scientists must assume determinism, at least for the specialty they are working in. Religious folks must assume indeterminism as support for their belief in free will and the possibility of surviving as a non-physical entity after dying. Both opposing philosophical positions are highly ingrained, and would remain so except for one thing: Humans, being highly curious and capable, continue to broaden their experience with the external world. Again, determinism allows us to manipulate our surroundings, increasing life spans and the enjoyment thereof. Indeterminism, being a failed assumption with no data in support, has added nothing to our understanding and enjoyment of the world around us.

DISINCARNATION. The convenient religious belief that the spirit leaves the body upon death.

DOPPLER EFFECT. The shortening or lengthening of wave motion due to a change in the distance between the source and the observer. The Doppler Effect is observed whenever a train passes by as it whistles. When it is coming toward us, the pitch is high (short wavelengths); when it is going away from us, the pitch is low (long wavelengths). Thus, when the source is moving away, subsequent 1-second beats will take slightly longer to arrive at your ears because the sound produced has to travel a greater distance through the air than it would if the source was stationary.

$E=mc^2$. The equation promoted by and improperly interpreted by Einstein. It was originally devised by James Clerk Maxwell in 1862[125] and not attributed to him by Einstein until late in life.

EINSTEINISM. "A statement or prediction that is true, but for the wrong reason."[126] Other, less preferred definitions are: 1) "a joke that becomes much less funny if it requires an explanation."[127] 2) "the perturbation of language or perception in order to put a positive spin on some aspect of Einstein's life. It may include distortion, omission, falsification, or corruption of the historic record in order to promote Einstein."[128] The most disparaging view was presented by Schlafly: "It is all a myth. Einstein did not invent relativity or most of the other things for which he is credited. He is mainly famous for popularizing the discoveries of others. We have all been duped."[129] That's pretty harsh. Much of the math in relativity works just fine even though the interpretations are grossly incorrect.

ETHER. The proper spelling of the hypothesized fixed medium once thought to be responsible for light transmission.[130] Its existence was disproven by the Michelson-Morley Experiment.[131] Earth travels around the Sun at a velocity of 30 km/s. If the ether was fixed, as assumed, then Earth's motion relative to that fixed medium would have been 30 km/s. It was much less than that, proving that the fixed ether did not exist. In other words, the fixed "ether" was falsified, but the "aether" consisting of an "aetherosphere" around Earth was not.

EXISTENCE. The XYZ portion of the universe occupied by a microcosm after its formation via submicrocosmic convergence and before its destruction via submicrocosmic divergence. All microcosms contain matter and have mass.

---

[125] Ricker, 2015, The origin of the equation $E=mc^2$.
[126] http://go.glennborchardt.com/Borchardt14massenergy
[127] Attributed to Ian James Hay at http://go.glennborchardt.com/einsteinism
[128] Moody, 2009, The eclipse data from 1919: The greatest hoax in 20th century science.
[129] Schlafly, 2011, How Einstein ruined physics.
[130] As defined by Farmer, 1997, Universe alternatives. The word I now use for the medium for light transmission, gravitation, and the formation of baryonic matter is aether, which is not fixed.
[131] Michelson, A.A., and Morley, E.W., 1887, On the relative motion of the earth and the luminiferous ether. [Often referred to as "MMX."]

FINITE PARTICLE THEORY (FPT). Indeterministic theory based on the assumption of microcosmic finity. Proponents tend to be absolutists who believe that the ideal end members of the matter-space continuum actually exist. The earliest version was proposed by the Greek atomists, who thought that matter consisted of perfectly identical little balls filled with "solid matter," in opposition to Aristotle's view that matter was infinitely subdividable. Accelerator experiments have supported Aristotle, but there still are believers in FPT who assume that a solid subatomic particle eventually will be found.[132] In this light, the Higgs boson has been anointed by the mass media as the "god particle." It supposedly is responsible for all the mass in the universe, although, ironically, it exists outside, not inside ordinary matter.

GENERAL RELATIVITY THEORY (GRT). The 1916 follow-up to Einstein's Special Relativity Theory[133] that continued his 1905 objectification of time, combining it with matter to propose space-time as a 4-dimensional object. GRT is the essential foundation of the Big Bang Theory. Like Special Relativity Theory, it includes numerous ad hoc assumptions that violate earlier laws of physics that form essential parts of classical mechanics.

GRAVITATION. The tendency for microcosms to be pushed toward less active, more massive microcosms.

HEISENBERG UNCERTAINTY PRINCIPLE. The claim by Werner Heisenberg that it is impossible to determine both the velocity and position of a particle at the same time. This marked the death of finite universal causality common to Newtonian mechanics, implying instead that causality was infinite. See Copenhagen Interpretation for the regressive circumvention of this calamity with respect to quantum mechanics.

IDEALISM. The philosophical tendency to extend imagined things and events into the real world.

---

[132] Abu-Bakr, 2007, The End of Pseudo-Science.
[133] Einstein, 1916, The foundation of the general theory of relativity; Einstein, 1916, Hamilton's principle and the general theory of relativity; Einstein, 1917, Cosmological considerations of the general theory of relativity.

IMMATERIALISM. The solipsistic belief that the universe does not consist of matter in motion. In opposition to materialism, immaterialism assumes that the universe is an illusion or internal perception that would not exist without a perceiving being.

INPERIMENT. A thought "experiment." I invented this term as a proper replacement for what was formerly considered a "thought experiment" by quasi-immaterialists such as Einstein. Strictly speaking, an experiment only can occur outside the mind per the prefix "ex." Science discovers truth through observation and experiment. Inperiments may be useful for predicting experimental results, but they have little credence among materialists (scientists) until those experiments actually are performed.

INDETERMINISM. The assumption that some effects may not have material causes (e.g., ESP, ghosts, gods, creation, free will). Specifically, indeterminism is founded on fundamental assumptions that lead to the false conclusion that free will is possible. Attempts to provide scientific support for indeterminism are inherently in contradiction. Science only deals with matter in motion; imagined "things" that do not exist cannot be tested because they do not exist.

LORENTZ CORRECTION FACTOR, γ. An equation derived by physicist Hendrik Lorentz, which is used throughout the mathematics of relativity:

$$\gamma = \frac{1}{\sqrt{1 - \frac{v^2}{c^2}}}$$

Unlike others, I call it a "correction" factor, because that is all it is. Wave motion through a medium takes time. For instance, when source and detector are moving apart, the path of that motion is stretched out: it takes longer to detect that motion. One needs to know the distance between the source and observer. Wikipedia plainly shows it to be nothing but an uncomplicated *measurement* problem:[134]

---

[134] http://go.glennborchardt.com/TimeDilation

Simple inference of velocity time dilation [edit]

Time dilation can be inferred from the observed constancy of the speed of light in all reference frames dictated by the second postulate of special relativity.[12][13][14][15]

This constancy of the speed of light means that, counter to intuition, speeds of material objects and light are not additive. It is not possible to make the speed of light appear greater by moving towards or away from the light source.

Consider then, a simple clock consisting of two mirrors A and B, between which a light pulse is bouncing. The separation of the mirrors is L and the clock ticks once each time the light pulse hits either of the mirrors.

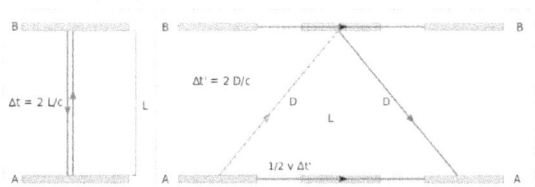

Left: Observer at rest measures time $2L/c$ between co-local events of light signal generation at A and arrival at A
Right: Events according to an observer moving to the left of the setup: bottom mirror A when signal is generated at time $t'=0$, top mirror B when signal gets reflected at time $t'=D/c$, bottom mirror A when signal returns at time $t'=2D/c$

In the frame in which the clock is at rest (diagram on the left), the light pulse traces out a path of length $2L$ and the period of the clock is $2L$ divided by the speed of light:

$$\Delta t = \frac{2L}{c}.$$

From the frame of reference of a moving observer traveling at the speed v relative to the resting frame of the clock (diagram at right), the light pulse is seen as tracing out a longer, angled path. Keeping the speed of light constant for all inertial observers, requires a lengthening of the period of this clock from the moving observer's perspective. That is to say, in a frame moving relative to the local clock, this clock will appear to be running more slowly. Straightforward application of the Pythagorean theorem leads to the well-known prediction of special relativity:

The total time for the light pulse to trace its path is given by

$$\Delta t' = \frac{2D}{c}.$$

The length of the half path can be calculated as a function of known quantities as

$$D = \sqrt{\left(\frac{1}{2}v\Delta t'\right)^2 + L^2}.$$

Elimination of the variables D and L from these three equations results in

$$\Delta t' = \frac{\Delta t}{\sqrt{1-\frac{v^2}{c^2}}},$$

which expresses the fact that the moving observer's period of the clock $\Delta t'$ is longer than the period Δt in the frame of the clock itself.

In addition to the velocity of light, the "*c*" in the equation can be replaced by the velocity inherent to wave motion in any particular medium. For instance, sound in the atmosphere at sea level usually travels at 343 m/s at 20°C. In all media, changes in the travel path between source and detector produce the Doppler Effect. The phenomenon that regressives call "time dilation" occurs when source or detector move away from each other and I suppose they might call it "time contraction" when they move toward each other. In the first instance, one must multiply by γ to calculate the increased length of the wave emitted from the source. In the second, one must divide by γ to calculate the decreased length of the wave emitted from the source. Note also that if *c* were infinite, γ would be 1 and there would be no effect.

As seen from the above, the Lorentz correction factor, γ, only concerns *measurement*. Strictly speaking, it has nothing to do with reality and everything to do with how we measure that reality. Time does not really dilate; mass does not really increase; length does not really contract as a result of motion through the imagined perfectly empty space. Claims to the contrary, such as those engendered from indeterministic interpretations of Special Relativity Theory, have no merit aside from unrecognized univironmental interactions that may occur in a special milieu that is by no means perfectly empty as assumed by Einstein and his followers who misinterpret the Lorentz Correction Factor.

MACROCOSM. The environment of a microcosm. Strictly speaking, the macrocosm contains the rest of the infinite universe. Practically speaking, only the nearby portions of the universe generally have much influence on a particular microcosm.

MASS. Resistance to acceleration. We determine the mass of a microcosm by accelerating it with another microcosm of known mass and velocity. Because all microcosms in the universe are in motion, mass and velocity are relative, so we have established standards, which on Earth, are relative to the acceleration of gravity (9.81 m/s^2). Because length and time also are relative, we have established conventions for those too. Realize, however, that these conventions are not absolute. Because the universe is infinite, measurements for each of them have a plus or minus and each tends to change over time and location. That is why we occasionally add a leap second to the length of the day as Earth's rotation rate slows. Although related, mass and matter are not identical. Mass is dependent on the internal motion of submicrocosms, which increases with the absorption of motion and decreases with the emission of motion per Maxwell's $E=mc^2$ equation.[135]

---

[135] Borchardt, 2009, The physical meaning of $E=mc^2$.

MATERIALISM. The assumption that the external world exists after the observer does not, and that the universe consists of matter. It is the First Assumption of Science.

MATTER. An abstraction for all things in existence. Above all, matter always contains other things within and without, ad infinitum. There are two basic types of matter: baryonic and aether. Although baryonic matter is what we ordinarily observe, aether is tiny and normally not directly detectable. Both have mass produced by constituents subject to interactions demonstrated by the $E=mc^2$ equation. Both are portions of the universe and have three XYZ dimensions. The "solid matter" of the idealist does not exist.

MATTER-MOTION TERMS. Valuable calculations in physics that generally multiply terms for matter and terms for motion. These describe neither matter nor motion, although they commonly have been objectified in regressive physics. For more details, see the chapter on progressive physics in "Infinite Universe Theory."

MATTER-SPACE CONTINUUM. A range or series of microcosms that are slightly different from each other and that exist between what we imagine to be perfectly solid matter and perfectly empty space.[136] Like all idealizations, perfectly solid matter and perfectly empty space do not and cannot exist.

The matter end member:

As mentioned, matter is an abstraction; there is no such thing as matter per se—there are only individual, unique examples of matter. As an abstraction, matter is akin to "fruit"—there are only individual, unique examples of fruit. The idea that solid matter must exist deep down at some level is still just that, an idea, ideal, or idealization that cannot occur in nature. The Greek atomists imagined that atoms were true elementary particles filled with solid matter. The things we now call atoms appear to contain mostly empty space. Even so, some absolutists assume that we just

---

[136] Modified from Merriam-Webster: http://go.glennborchardt.com/continuum

have not gone far enough and that the nirvana of perfect solidity is theoretically possible.[137] At one time, the space between you and me may have been considered empty. Now we know that is not the case, for space is just the stuff that yields to the motion of other stuff. These ideals exist only in our brains—they help us understand the properties of various kinds of matter, but they can have no real existence. We use them to understand the intervening reality. It is good enough for finding a doorway instead of a wall, even though the doorway contains matter in the form of air (and aether) and the wall contains space. In Infinite Universe Theory, what we consider solid matter is simply a portion of the universe that offers more resistance to acceleration than other portions we consider empty space.

The space end member:

The absolutist's belief in the ideals of perfectly empty space, nothing, and nonexistence comes right out of the cosmogonical handbook whose precursors are the sacred texts of traditional religion. To insist, like the young Einstein and his positivist friends, that space is perfectly empty or immaterial makes one a rank idealist. To insist, as indeterminists are wont to do, that idealities could be or must be realities merely provides another roadblock to the ultimate acceptance of Infinite Universe Theory. The antidote begins with similarity analysis, which helps us tell the difference between what is ideal and what is real.

MECHANICS. The study of the universe in terms of matter and the motion of matter.

MECHANISM. The philosophy based on mechanics, which asserts that the universe has only two phenomena: matter and the motion of matter. The term also is used more widely to describe the physical interactions that produce a particular effect.

MECHANIST. One who assumes that the universe consists only of matter in motion.

---

[137] Abu-Bakr, 2007, The End of Pseudo-Science.

Glenn Borchardt

MICROCOSM. An XYZ portion of the universe surrounded by an equally important environment called a macrocosm. Note that in conventional science microcosms are referred to as systems, which generally are considered more important than the environments in which they exist. In Infinite Universe Theory, microcosms cannot exist without their equally important macrocosms. Regardless of the immensity of a microcosm, in an Infinite Universe an infinitely large macrocosm still surrounds it. The boundaries of a system sometimes are obvious: An apple, for instance, has a skin that roughly distinguishes it from its surroundings. At other times, the boundaries are not so obvious: A bee colony, for instance, has rather obscure boundaries when many of its members are far afield gathering nectar. Boundary selection is often difficult, always important, and frequently arbitrary. As scientists, we try to reduce arbitrariness by recording the location of boundaries with as much accuracy as possible. Our designation of a particular XYZ portion of the universe as a microcosm faces the same problems, although in that instance, we treat its environment (the associated macrocosm) as equally important. Also, by attempting to treat the microcosm and the macrocosm equally, we are not as likely to miss important factors, as we would if we were biased toward one or the other.

MIRACLE. An impossibility occurring only in dreams or imagination, often retold in religious books and sermons via the telephone chain effect. Rare events, such as coincidences, which occur in the real world, are not miracles.

MOTION. An abstraction for all occurrences.

NOMA. Nonoverlapping magisteria, a concept invented by the late agnostic, Stephen Jay Gould, who proposed that science and religion are merely two different ways of looking at the same thing.[138] Science is objective and religion is subjective. Science describes objects and predicts what those objects will do; religion proclaims morals and values. He thought a truce between the two opposed views would arise if each remained on its own turf.

---

[138] Gould, 1997, Nonoverlapping magisteria.

NONES. In the US, the newly coined term for the religiously unaffiliated population currently increasing in numbers, particularly among the young.[139] Most are just skeptical, with a few calling themselves agnostics, and fewer still admitting to being atheists.

OBJECTIFICATION. The consideration of motion as an object. Humans have objectified motion throughout history. Today, objectification is one of the primary roadblocks to understanding Infinite Universe Theory. Of the two fundamental phenomena, matter often comes to mind before motion. You can see matter, but you cannot see motion as a thing apart from matter. As we have seen, Einstein's relativity reeks with this old-fashioned tendency to objectify motion.[140] His "corpuscular" theory of light has a historic parallel with the "caloric fluid" theory of heat. Antoine Lavoisier (1743-1794),[141] the father of chemistry, imagined that "caloric fluid" traveled from object-to-object as a material entity. When you touched a burning ember, the caloric fluid supposedly flowed into your finger. Of course, what was once considered a "thing" is now correctly considered "motion," the vibration of things. In other words, heat is motion, not matter. If you wait too long to remove your finger the vibrations in the ember stimulate vibration in your finger, causing your discomfort and serious chemical transformations in your skin. Strictly speaking, heat does not exist; it occurs. In this case, the only thing that exists is the vibrating molecule. The molecule takes up three dimensions, while its motion does not.

Physicists have long abandoned the notion that heat is matter, although they have not done so for light. Today, few would ask whether heat had mass, since even today's most regressive physicists must regard heat as motion. Likewise, waves do not have mass, although the medium through which they travel does.

Even when Einstein made odious mistakes in math, indeterminists looked the other way. Steven Bryant[142] and I traced this objectification to

---

[139] Pew Research Center, In U.S., Decline of Christianity Continues at Rapid Pace.
[140] Borchardt, 2011, Einstein's Most Important Philosophical Error.
[141] Note that Newton thought light to be a corpuscle as well.

Einstein's mathematical somersault in which he first properly derived l (length) from *c*t (velocity multiplied by time is always distance).[143] Unfortunately, by a little sleight of hand, he then used l, length, as a replacement for t, time. That is how time got to be a dimension. In a way, this has been sanctified in our conventional use of the term "light year," which is the distance light travels in a year. This is an extremely valuable "distance" measurement, but it does not make time a distance.

The General Theory of Relativity reflects this tendency to objectify motion on the grand scale. Time has no dimensions and is not "part" of the universe, although time occurs within the universe. Of course, the concept of "space-time" purports to "combine" space and time. However, only things can be "combined." "Time," having neither three dimensions nor existence, cannot be combined with anything. True, in our heads we can combine concepts, ideas, stories, and equations about real things and real motions, but that does not give them dimensions. A picture of a running dog is not a dog.

"Space" exists, but "space-time" does not. Nonetheless, it is commonplace for cosmogonists to assume that 4-dimensional space-time actually exists. As mentioned previously, this belief is required for the expanding universe hypothesis. Without the mathematical fabrication of space-time, the Big Bang Theory would not have survived as long as it did.

Why has Einstein's objectification of motion been so popular and enduring? I hinted at the reasons for that in my explanation of the deterministic assumption of ***inseparability***. In particular, the belief in motion without matter has been well-established ever since the first humans tried to understand the wind in the willows. That must have been rather frightening. They could not have known that air consisted of unseen particles now called nitrogen and oxygen. Here was a "thing" that was not a thing, like the proverbial ghost that was a thing, but not a thing. The imagined ghost was capable of traveling through walls. It supposedly had

---

[142] Brilliant mathematician who forgoed studies toward a Berkeley Ph.D. in physics due to what he recognized early as the regressive nature of the curriculum.
[143] Bryant and Borchardt, 2011, Failure of the relativistic hypercone derivation.

three dimensions and location, but certainly was not material. The unseen, unseeable causes of motions were given names—with a god for wind and a god for thunder, and on and on.

The indeterministic vestiges of the idea of matterless motion are nearly as dominant now as they were in 1905. Matterless motion always has been a mainstay of religion, from holy ghosts, to souls, to gods. Most folks talk about such "things" as if they existed. Thus, it was not surprising that Einstein and many others would objectify time to great applause. Modern physics is founded, not on the assumption of ***inseparability***, but on its indeterministic opposite, *separability*. Einstein and followers never understood this. Make no mistake about it. Time cannot dilate and space-time does not exist. Many of the paradoxes and absurdities in modern physics and cosmogony are traceable to this single most critical philosophical error. We can do better, but only if we give up the idea of matterless motion.

PARADIGM. A Greek word expressing a pattern of thought that generally includes a prominent theory and its associated corollaries used to interpret certain aspects of the world. With respect to science, Thomas Kuhn defined a paradigm as "universally recognized scientific achievements that, for a time, provide model problems and solutions for a community of practitioners."[144] Examples include neo-Darwinian evolution, plate tectonics, relativity, and the Big Bang Theory. Other terms for it include: "the conventional wisdom" or "mainstream thought." Practitioners closely guard the paradigm, usually rejecting non-compliant papers and proposals for employment or financial support that do not conform. Work within a paradigm is considered "normal science," whereas largely unsupported work outside it is usually considered "crackpot," if unsuccessful or "revolutionary," if successful. As with all revolutions, the overthrow of one paradigm by another is considered as difficult as it is monumental.

PARADOX. "Also known as an antinomy, is a logically self-contradictory statement or a statement that runs contrary to one's expectation. It is a statement that, despite apparently valid reasoning from true premises, leads

---

[144] Kuhn, 1996, The structure of scientific revolutions, p. 10.

to a seemingly self-contradictory or a logically unacceptable conclusion."[145] At least one of the supposed "true" premises actually is false.

POSITIVISM. A branch of the philosophy developed from empiricism that assumes that things unsensed do not exist. One variant, operationalism, assumes that things not detected through some mechanical operation do not exist. Thus, positivists reject theism, metaphysics, and speculation that they regard as having no basis in prior experience. They generally assume that space is perfectly empty despite the lack of evidence for such and the theoretical necessity for the existence of aether.

Like classical mechanics and classical determinism, positivism was initially of great value in combating indeterminism. It is still used today when atheists assert that a particular god does not exist because they cannot find evidence for it. Unfortunately, this approach tends to fail with advances in instrumentation. We cannot see the nitrogen, oxygen, argon, and carbon dioxide in air, but now we can detect them just fine with instruments. Today, positivism is found mostly on the indeterministic side of the philosophical struggle. For instance, religious positivists use the "god of the gaps" argument to claim that missing transitional fossils in parts of the sedimentary record are proof that evolution is false. And then, of course, aether denial is nothing, if not, positivistic.

POSTDICT. The opposite of "predict." Postdiction is the attempt to determine past events via a model generally used for predicting future events.

PROGRESSIVE PHYSICS. The deterministic version of physics that will replace the regressive version that became dominant with the increasing popularity of Einstein's relativity. There always have been regressive elements within physics, but the obvious paradoxes and contradictions brought forth by relativity now bring the regression to the fore. Matter-motion terms eventually will be seen as what they are: calculations, not as things or motions. Progressive physicists assume *infinity*, believe that light is a wave, that aether exists, and that momentum, force, energy, and space-

---

[145] Wikipedia.

time do not. Most do not believe Einstein's Untired Light Theory and that the universe is expanding after beginning via an explosion from nothing in violation of *conservation*.

REDSHIFT. The process by which wavelength becomes longer and/or frequency decreases. The "red" in the term implies an increase in wavelength derived from the fact that low-frequency red light has a longer wavelength than high-frequency blue light. It is a bit of a misnomer because color is determined by frequency, not wavelength. That is why half the distant galaxies actually are not red. There are many different types of redshift as explained in the text. When wave velocity is constant, the relationship: $v = \lambda f$ remains constant, where v = velocity, m/s, λ = wavelength, m, and f = frequency, cycles/s. There are many types of redshift, which are discussed in "Infinite Universe Theory."

REFORMIST PHYSICS. Any attempt to resolve the paradoxes and contradictions of modern, regressive physics by modifying relativity without discarding its indeterministic/religious assumptions entirely. Ever since relativity became popular, thousands of mostly unfunded skeptics have voiced objections and proposed modest and sometimes immodest alternatives. Its overthrow is the grand prize among those who view its paradoxes and contradictions as signs that relativity is ripe for the picking. In 2012, the Jean de Climont group in France developed a list of more than 8,000 dissident scientists with some presence on the Internet who objected to relativity and quantum mechanics.[146] The 2020 version includes over 9,500 dissidents and over 2,500 alternative theories, which the group says are "all amazingly very different."[147] There are over 550 alternative theories that use aether alone. Bet you never heard of any of these. Can you see why the media tends to shy away from any one of them? That is like the legal attack that dumps truckloads of irrelevant documents on the defense lawyers. Of course, reporters with even a smidgeon of knowledge about physics and cosmology are rare. The knowledgeable ones need to defend mainstream theories that they have already promoted. Also, in the

---

[146] de Climont, 2012, The Worldwide List of Dissident Scientists.
[147] de Climont, 2020, The worldwide list of alternative theories and critics. [The list has about 50% scientists.]

interest of sales, reporters must confirm the views of their audience. Efforts to destroy those views will not be met with open arms. In any case, reporters do not have the time or interest to sort through hundreds of alternatives to what they firmly believe anyway.

Reform, of course, is not up to the media. In physics and cosmology, the switch from one paradigm to another is the job of scientists who are not physicists and cosmologists. Nonetheless, as in the free will debate, the reform discussions currently are interminable. One wag even summed it up with something akin to the Second Law of Thermodynamics: "Discussions about Special Relativity naturally and quickly degrade into disorder and nonsense." Nonetheless, folks continue to seek compromises that might leave enough of relativity and cosmogony to be acceptable to the mainstream. Above all, one must be able to understand the numerous einsteinisms in which relativity got the right answers for the wrong reasons. In other cases, Einstein's interpretations, such as the Untired Light Theory, are just plain wrong.

Except for the dissident press, manuscripts unfavorable to relativity or the Big Bang Theory normally get the circular file. Partly this is because much dissident work is dreadful; not amounting to much more than the kind of silly modifications suggested by funded practitioners. Some of it is overtly religious or entertains other outrageous propositions. I have attended dissident talks proclaiming that the biblical flood covered most of the western US and formed the Grand Canyon. A few still insist that the Sun revolves around the Earth. One prevalent complaint about the dissident community is that the members seldom cite one another and that there is little co-authorship and that fundamental assumptions are rarely stated. Nonetheless, there has been much fine work done by a select few dissidents, Sagnac, for instance. Einstein has been proven wrong more often than he has been "proven right." These instances receive little publicity from the popular, indeterministically flavored press. As always, the main problem with reform is that it does not go far enough. As with agnosticism generally, mixing progressive elements with regressive elements will not remove the contradictions in interpretation.

REGRESSIVE PHYSICS. The indeterministic version of physics otherwise known as "modern physics." It describes the radical 20th-century departure from determinism that became popular after Einstein's introduction of Special Relativity Theory in 1905. Today the paradigm still has a solid grip on physics despite being plagued by numerous paradoxes and contradictions in the interpretation. Practitioners are heavily financed and allowed to publish outlandish deductions and wild speculations as long as they do not contradict relativity and cosmogony. Among these are the concepts of massless particles, immaterial fields, wormholes, multiverses, space-time and other fabrications, such as string theory that claim anywhere between one and 26 dimensions. Regressive physicists generally do not know what time is and that it cannot dilate. They believe that momentum, force, energy, and space-time actually exist, and that aether does not. Most do not know the proper interpretation of Maxwell's $E=mc^2$ equation. Regressives unwittingly accept Einstein's Untired Light Theory and the resulting interpretation that the universe is expanding and had a beginning via explosion from nothing. Today, almost all gainfully employed physicists appear to be regressive, with dissenters having been weeded out long ago.[148]

The regression began in response to the deterministic ravages of classical mechanism and dialectical materialism during the late 19th century (Darwin, Marx, etc.). In particular, Lenin's "Materialism and Empirio-criticism"[149] and the rise of communism put the fear of god into the west. Relativity's immaterialistic assumptions and paradoxes fit in with the religious beliefs of the day. Einstein prepared the way for the expansionist movement and the priest, Lemaître, suggested this meant the entire universe exploded from a "cosmic egg" in tune with Genesis. To get popular, all ideas, theories, papers, and books must fit the macrocosm in which they exist. Each acts as a weapon in the philosophical struggle between determinism and indeterminism, which continually involves spiralic progress through education and regress through miseducation. When first introduced, the absurdities in Special Relativity Theory and General Relativity Theory brought forth numerous complaints, but these were dismissed in favor of the financial rewards available to Einstein's supporters. When viewed with indeterministic eyes, falsifications of

---

[148] http://go.glennborchardt.com/Borchardt12censorship
[149] Lenin, 1908, Materialism and Empirio-criticism.

relativity were ignored, while so-called confirmations were revered. Opponents usually were dismissed as "cranks" or "crackpots" no matter how logical their arguments may have been.

What makes relativity so complicated is that some of the predictions have been confirmed even though the interpretations may be incorrect. The challenge for progressive physics is to find the actual physical causes for those confirmations.

SECOND LAW OF THERMODYNAMICS (SLT). "The Second Law of Thermodynamics states that all isolated systems eventually run down. In the regressive interpretation, constituent matter supposedly is converted into "energy," which escapes the isolated system as unusable heat. Another way of stating it from a mechanical viewpoint is that the constituents of the system will diverge or expand into its surroundings via their own momenta. Either way, both interpretations fit the expanding universe of the Big Bang Theory with divergence being assumed greater than convergence."[150]

SIMILARITY ANALYSIS. In the early 1970's I invented the SIMAN coefficient for comparing multivariate analyses of neutron-activated samples of volcanic ash.[151] It is a simple mathematical formula nicely illustrating the matter-space continuum in which perfect similarity yields a coefficient of 1 and perfect dissimilarity yields a coefficient of 0. No real samples ever have those values, but values over 0.95 generally indicate volcanic ashes from the same eruption. The coefficient is used worldwide, with the 1972 paper being one of our most cited. It may have influenced my later discovery of the Ninth Assumption of Science, *relativism*.

SOLIPSISM. The self-centered belief that the existence of the universe depends on consciousness. Solipsists are immaterialists, who in the extreme might even think that the entire external world depends on them and then disappears when they do. Einstein was being solipsistic when he

---

[150] Borchardt, 2017, Infinite Universe Theory.
[151] Borchardt, 1974, The SIMAN coefficient for similarity analysis; Borchardt and others, 1972, Correlation of the Bishop Ash, a Pleistocene marker bed, using instrumental neutron activation analysis.

claimed that fields were "immaterial." Solipsism is the indeterministic opposite of materialism.

SPACE. An abstraction for matter that has less mass than its surroundings. As with all abstractions, there is no such thing as "space," there only are individual examples of space, such as what lies between the electron and the proton, between doorposts, or between galaxies. There can be no "perfectly empty space" or absolute vacuum devoid of matter.

It might be helpful to think of "empty space" as though it were a scaled-down Milky Way. No matter how small the scale, there is always some matter (like the stars) separated by what at first might seem to us as "empty space" (like the interstellar regions). This is the essence of the consupponible assumptions of infinity (The universe is infinite, both in the microcosmic and macrocosmic directions) and interconnection (All things are interconnected, that is, between any two objects exist other objects that transmit matter and motion). As mentioned many times before, in Infinite Universe Theory we realize that "perfectly solid matter" and "perfectly empty space" are only idealizations. The reality always is something in between. This implies that non-existence is impossible for each part of the universe, no matter how large or small. It is impossible for the Infinite Universe not to exist. The nothingness that indeterminists have imagined is only an idea. There are two things that the Infinite Universe cannot produce: perfectly solid matter and perfectly empty space.

Again, our slogan that "space is matter" is based on the observation that no portion of the universe is completely void of matter. We cannot produce a perfect vacuum and the 2.7°K Cosmic Microwave Background tells us that even intergalactic space contains microcosms in motion. The completely empty space assumed by Einstein and other aether deniers would have a temperature of 0°K. At the other end of the continuum, black holes, if they exist, could not contain "solid matter" without "empty space." They would be subject to Maxwell's $E=mc^2$ equation—they would emit radiation just like all other microcosms. Even Stephen Hawking finally admitted that this is true—they are not black, like he first assumed, but grey.[152]

---

[152] Lewis, 2014, Grey is the new black hole.

SPACE-TIME. A matter-motion term for an idealization or visualization of the location of things with respect to the past or future. Note that, like all matter-motion terms, space-time neither exists nor occurs; it is a concept.

SPECIAL RELATIVITY THEORY (SRT). Einstein's unattributed development of Maxwell's 1862 $E=mc^2$ equation in which he erroneously considered light to be a massless particle having constant velocity.[153] His theory of light required eight absurd ad hocs (Table 3). SRT makes plentiful use of the Lorentz Correction Factor, γ, which has been used to support solipsistic claims of time dilation, increased mass, and decreased length for objects in motion through empty space. SRT is a mixed bag, with the $E=mc^2$ equation being correct and many of the indeterministic claims being incorrect. For example, time, t, and length, l, are not interchangeable categories;[154] energy is not equivalent to mass, time cannot dilate, aether exists and photons do not, etc.

SUBMICROCOSM. A microcosm that exists inside another. A submicrocosm is always smaller than the microcosm that contains it.

SUPERMICROCOSM. A microcosm that exists outside another. A supermicrocosm can be either larger or smaller than the microcosm.

SYSTEMS PHILOSOPHY. The philosophy that treats portions of the universe as systems more important than their surroundings. That myopic view has been with us for millennia, but was only formally introduced by Ervin Laszlo in 1972.[155] It remains the current scientific world view, with the archetype of systems philosophy being the Big Bang Theory.

TELEPHONE CHAIN. Game in which a statement is whispered from one person to another in sequence. Due to imperfections in speaking and hearing, the end result of a long chain of participants often can be quite amusing. In religion it is not amusing, resulting in outrageous assertions claiming real-world support for miracles and other fantasies sometimes

---

[153] Einstein, 1905, On the electrodynamics of moving bodies.
[154] Bryant and Borchardt, 2011, Failure of the relativistic hypercone derivation.
[155] Laszlo, Ervin, 1972, Introduction to Systems Philosophy: Toward a New Paradigm of Contemporary Thought.

used to kill others. In science it has numerous analogies whenever microcosms or their motions are being reproduced. Examples include cell reproduction that produces strange mutations and imperfect wave reproduction that produces the cosmological redshift.

TEMPLETON FOUNDATION. "The John Templeton Foundation is a philanthropic organization that reflects the ideas of its founder, John Templeton, who became wealthy after a career as a contrarian investor and wanted to support progress in religious and spiritual knowledge, especially at the intersection of religion and science." (from templeton.org.) The foundation has a billion-dollar endowment and hands out over $60 million each year, with the over a million-dollar Templeton Prize being coveted among the religious. "Previous winners include Mother Teresa, Billy Graham, Aleksandr Solzhenitsyn, and Charles W. Colson, the born-again Watergate convict. Over the past 20 years, most of the winners have been scientists who see inklings of the divine in nature, including Paul Davies, Freeman J. Dyson, John C. Polkinghorne, and Charles Hard Townes. This year's winner, John D. Barrow, a cosmologist at the University of Cambridge, continues in that vein." (Horgan, 2006) According to Reuters: "Brazilian physicist and astronomer Marcelo Gleiser has been awarded the 2019 Templeton Prize, worth $1.4 million, for his work blending science and spirituality." The 2020 prize winner was the notorious "geneticist and physician Francis Collins, Director of the National Institutes of Health, who led the Human Genome Project to its successful completion in 2003 and throughout his career has advocated for the integration of faith and reason…"[156]

TIME. Motion. Universal time is the motion of each thing with respect to each of the other things in the Infinite Universe. Specific time is the motion of one thing with respect to another. All events and changes are motions. Clocks measure motion and would be impossible without it.

TIRED LIGHT THEORY (TLT). A theory based on Hubble's later suggestion that the universe was not necessarily expanding if the cosmological redshift was due to some unbeknownst cause in addition to

---

[156] https://go.glennborchardt.com/Collins-Temple20

the well-known Doppler Effect. In a fashion that was to become manic and knee-jerk, regressives rejected suggestions that did not use Einstein's religious assumption that space was perfectly empty.

UNIVIRONMENT. A microcosm and its macrocosm. A word coined by Elizabeth Patelke and myself to overcome the microcosmic overemphasis of systems philosophy (of which the finite universe is the archetype). It was first printed in the 1984 review manuscript of "The Scientific Worldview."[157] The unification also was a rejection of "environmental determinism," with its overemphasis on the macrocosm.

UNIVIRONMENTAL DETERMINISM. The philosophy and universal mechanism of evolution based on the observation that what happens to a portion of the universe is determined by the infinite matter in motion within and without. In addition to being the universal mechanism of evolution, univironmental determinism also was proclaimed to be "The Scientific Worldview" in the review manuscript in 1984[158] and the published book in 2007.[159]

UNTIRED LIGHT THEORY (ULT). The belief that, light, unlike other things or motions can travel from one point to another without losses, in contradiction of the Second Law of Thermodynamics. The opposing belief is called the "Tired Light Theory," which states that the cosmological redshift is due to imperfect reproduction of waves as they travel through the aether medium. ULT is required for the interpretation that the universe is expanding.

WOO-WOO or WOO. "Unconventional beliefs regarded as having little or no scientific basis, especially those relating to spirituality, mysticism, or alternative medicine." (Lexico)

---

[157] Borchardt, 1984, The Scientific Worldview.
[158] Ibid.
[159] Borchardt, 2007, The Scientific Worldview: Beyond Newton and Einstein.

Appendix

I sure hope you enjoyed reading *Religious Roots of Relativity* as much as I did in writing it.

It would be wonderful if you could give a review of it.

During the next few decades science will undergo its greatest paradigm shift as Big Bang Theory is replaced by Infinite Universe Theory. Join in the Last Cosmological Revolution by getting occasional emails concerning our progress toward that goal. There will be breaking news, links to the latest Blog entry, course announcements, and info on how you can get free books, so be sure to subscribe to the email list:

http://go.glennborchardt.com/emailsub

Glenn Borchardt

Dr. Glenn Borchardt has over sixty years of practical and theoretical experience in science and philosophy. He has produced over 500 scientific reports, including journal articles, books, chapters, abstracts, computer programs, and consulting reports. His most notable books are: "The Ten Assumptions of Science," which opposes the foundations of science and religion, "The Scientific Worldview," which proclaims the universal mechanism of evolution as the key to understanding the universe, and "Infinite Universe Theory," as the ultimate replacement for the Big Bang Theory. Borchardt is the Director of the Progressive Science Institute in Berkeley California.

Religious Roots of Relativity shows that, unlike other scientific theories, relativity is founded on religious assumptions. Glenn Borchardt, author of The Ten Assumptions of Science, elaborates on the opposing indeterministic assumptions to present "The Ten Assumptions of Religion" as the framework for this new book. Each fundamental religious assumption is shown to have much in common with the fundamental assumptions Einstein subconsciously used in devising Special and General Relativity Theory. One theme runs through the entire book: Einstein's erroneous assumption that space was perfectly empty. That was critical for his popular Untired Light Theory, as it has been for popular biblical creation stories, and for popular Big Bang Theory. There is no evidence, however, for perfectly empty space; it is only an idealization akin to the dreams and imaginings of religion. It cannot possibly exist. Nonexistence, nothingness, therefore is impossible. The universe exists everywhere and for all time. Without relativity and its foundation in religion, the book predicts Big Bang Theory will be victim to the Last Cosmological Revolution: Infinite Universe Theory.

This is the book for you if you have wondered why relativity has remained lucrative and popular despite its weird paradoxes, contradictions, and interpretations. This is the book showing the intimate, necessary connection between relativity and religion, which has led to relativity's longevity and indubitable veracity among those who still hold fast to religious assumptions.

Glenn Borchardt

"Wow! I finished reading your book in one day! I just couldn't stop scrolling the pages. It was an enjoyable read and very well written. You have a great writing style that is easy to read. Nice final sentence too." -Bill Howell

"Borchardt's new book is ultimately a fast read, because (like all his books) once you start reading it, you can't put it down. And, literally, you can't put it down physically, and you can't put it down argumentatively. Some may disagree with it. But that would only reveal the indeterminist within. Borchardt ends his masterpiece with a look forward to the inevitable paradigm shift, and how mankind will be better off for it." -Fred Frees

"Glenn Borchardt's book "Religious Roots of Relativity" is not just about relativity and religion, it's not only about physics, it's much more, about science which is under a siege by everything what is not science. If I had to review Borchardt's book: "Religious Roots of Relativity" in only one sentence, I would say: We need more books like this one!" -Rudolf Vrnoga

"Impressive piece of work! Very much in line with Collingwood and my essay on the subject. I had never realized these assumptions were of religious origin, though, besides the priest's obvious motivations." -Pierre Berrigan

Glenn Borchardt's book uses the hammer of Infinity to explain and destroy the junk theories that plague 'Official' physics today. This is a book that should be used in college courses, to give students a basic understanding of how physics is done. Physics has 'gone off the rails' for a century and it is books like Borchardt's that will return physics from its current unscientific and anti-materialist base and back on to a scientific and materialist road." -Mike Gimbel

Religious Roots of Relativity
Progressive Science Institute
20241211 9:04 AM

www.ingramcontent.com/pod-product-compliance
Lightning Source LLC
Chambersburg PA
CBHW070639220526
45466CB00001B/230